Mathematical Engineering

Today, the development of high-tech systems is unthinkable without mathematical modeling and analysis of system behavior. As such, many fields in the modern engineering sciences (e.g. control engineering, communications engineering, mechanical engineering, and robotics) call for sophisticated mathematical methods in order to solve the tasks at hand.

The series Mathematical Engineering presents new or heretofore little-known methods to support engineers in finding suitable answers to their questions, presenting those methods in such manner as to make them ideally comprehensible and applicable in practice.

Therefore, the primary focus is—without neglecting mathematical accuracy—on comprehensibility and real-world applicability.

To submit a proposal or request further information, please use the PDF Proposal Form or contact directly: *Dr. Jan-Philip Schmidt, Publishing Editor (jan-philip.schmidt@springer.com).*

More information about this series at http://www.springer.com/series/8445

Volker Simon · Bernhard Weigand
Hassan Gomaa

Dimensional Analysis for Engineers

Springer

Volker Simon
Progress-Werk Oberkirch AG
Offenburg
Germany

Bernhard Weigand
Institut für Thermodynamik der Luft- und Raumfahrt
Universität Stuttgart
Stuttgart
Germany

Hassan Gomaa
Large Gas Turbines
Siemens AG
Berlin
Germany

ISSN 2192-4732 ISSN 2192-4740 (electronic)
Mathematical Engineering
ISBN 978-3-319-52026-1 ISBN 978-3-319-52028-5 (eBook)
DOI 10.1007/978-3-319-52028-5

Library of Congress Control Number: 2016963646

Printed on acid-free paper

This Springer imprint is published by Springer Nature
The registered company is Springer International Publishing AG
The registered company address is: Gewerbestrasse 11, 6330 Cham, Switzerland

To our teachers

Preface

At all times, humans have been striving to understand the world that surrounds them and that they live in. We want to understand, why things are happening the way they do. We try to comprehend, we try to understand, we are yearning for insight. From insight we hope to gain foresight, and from understanding we hope to be able to affect and manage our future.

We want to manage for the better, so we need to measure the status before and after we have taken action. We want to make sure by measurement that we have done the right thing. But what is the right measure?

Obviously, the fact that we gain insight is an advantage for us. And the event, when we gain insight, is pleasurable and satisfying for us: "Eureka!" When, after having worked hard and long to solve a difficult problem, after doubt and despair, after struggling to comprehend, when we finally understand: This moment is a most beautiful sensation.

Dimensional analysis is a method that helps to gain insight and that helps to understand physical problems. It is universally valid; it can always be applied; it is always correct. Dimensional analysis shows us the right measure, the correct scaling of a problem, which is always inherent to the problem itself.

This book aims to introduce the reader to the fascinating topic of dimensional analysis through a large number of examples and just as many fundamentals, which are needed to understand the working principles and foundations of the method. The book is organized in four chapters. The first chapter concerns the fundamentals of the method, terms and definitions and the core of the method: the Buckingham- or Pi-Theorem. The second chapter on model theory underlines its importance through brevity. The third chapter collects examples ranging from mechanics, fluid mechanics, thermodynamics, electrodynamics to a variety of physical settings to show the broad range of applications. In the fourth chapter, finally, the application of dimensional analysis to partial differential equations of physical problems is demonstrated, and the power of the method is best shown by concentrating on selected examples, the so-called similarity solutions. The appendix collects some exercises and corresponding solutions for tutorials and self-study.

We, the authors, independently enjoyed our first encounter with dimensional analysis as a universal method that helps understanding common phenomena, physical problems, and through understanding helps to solve engineering tasks at hand. And we certainly appreciated the skills of the teachers, who first showed and taught us the beauty of dimensional analysis. Two of us (Volker Simon and Bernhard Weigand) have greatly benefitted from our teacher Prof. Joseph H. Spurk, to whom we are grateful and express our thankfulness. His lectures and book on dimensional analysis have certainly strongly influenced our work and the way to look at problems and find their solutions.

We hope that our book has the same effect on its readers: It gives them insight into dimensional analysis, supports their understanding of the physical world, and as a consequence the method will be a pleasurable benefit to them.

Many people helped us in all phases of the preparation of this book. We thank very much Roman Frank who helped us with the figures. Many thanks also go to our students for discussions concerning the examples. Finally, we would like to express our thanks to Springer Press for the very good cooperation during the preparation of this manuscript.

Offenburg, Germany Volker Simon
Stuttgart, Germany Bernhard Weigand
Berlin, Germany Hassan Gomaa

Contents

Symbols

a	Acceleration (m/s^2)
a	Laplacian length (m)
a	Strain rate (1/s)
a_{ij}	dimension matrix (–)
a_s	Velocity of sound (m/s)
A	Area (m^2)
$\mathbf{B}$	Magnetic field (T)
c	Specific heat capacity (J/(kg K))
c	Wave speed (m/s)
C	Dimensional constant (–)
C	Capacitance (F)
d	Number of dimensionless variables (–)
d	Diameter (m)
D	Diffusion coefficient (m^2/s)
D	Diameter (m)
$\mathbf{D}$	Electric displacement field (C/m^2)
E	Energy (J)
E	Modulus of elasticity (N/m^2)
$\mathbf{E}$	Electric field (N/C)
f	Frequency (1/s)
f	Function (–)
F	Force (N)
F	Function (–)
g	Gravitational acceleration (m/s^2)
G	Gravitational constant (m^3/(kg s^2))
h	Heat transfer coefficient (W/(m^2 K))
h	Height, depth (m)
H	Height (m)
H	Enthalpy flux (m^3 K/s)
$\mathbf{H}$	Magnetizing field (A/m)

$\mathbf{I}$	Electric current (A)
J	Second moment of area (m^4)
$\mathbf{J}$	Electric current density (A/m^2)
k	Thermal conductivity (W/(m K))
k_B	Boltzmann constant (J/K)
k_j	Dimension exponent (–)
K	Kinematic momentum (m^4/s^2)
L	Length (m)
L	Inductivity (H)
m	Number of base quantities (–)
m	Mass (kg)
M	Shaft torque (Nm)
M	Scale factor (–)
M	Molecular weight (kg)
M	Planet mass (kg)
n	Volume specific number ($1/m^3$)
n	Natural frequency (1/s)
n	Rotational speed (1/s)
n	Number of physical quantities (–)
p	Physical quantity
p	Pressure (N/m^2)
P	Power (W)
P	Force (N)
q	Electric charge (C)
$\dot{Q}$	Heat flux (W)
r	Radial distance (m)
r	Rank of dimension matrix (–)
R	Radius (m)
R	Resistance (Ω)
R	Gas constant (J/(kg K))
R_{m}	Universal gas constant (J/(mol K))
t	Time (s)
t	Thickness (m)
T	Temperature (K)
u,v,w	Velocity (m/s)
U,V	Velocity (m/s)
v	Specific volume (m^3/kg)
v	Fluctuation velocity (m/s)
$\mathbf{v}$	Velocity vector (m/s)
$\dot{V}$	Volume flow (m^3/s)
w	Deformation (m)

W	Resistance force (N)
x	Numerical value of physical quantity (–)
x,y,z	Coordinates (m)
X	Base unit (–)

Greek Letter Symbols

α	Thermal diffusivity (m^2/s)
α	Angle (–)
α	Scale factor (–)
β	Thermal expansion coefficient (1/K)
γ	Angle of attack (–)
Γ	Circulation (m^2/s)
δ	Boundary layer thickness (m)
ε	Restitution coefficient (–)
ε	Permittivity (F/m)
η	Dynamic viscosity (kg/(m s))
η	Non-dimensional coordinate (–)
ϑ	Temperature (K)
θ	Temperature (K)
λ	Wave length (m)
μ	Volume fraction ratio (–)
μ	Permeability (N/A^2)
ν	Kinematic viscosity (m^2/s)
Π	Dimensionless product (–)
ρ	Density (kg/m^3)
ρ	Volumetric charge density (C/m^3)
σ	Surface tension (N/m)
σ	Strain (N/m^2)
σ	Electric conductivity (S/m)
ϕ	Electric potential (V)
Ψ	Stream function (m^2/s)
ω	Angular velocity (1/s)
Ω	Rotational speed (1/s)

Subscripts

0	Initial state
0	In vacuum
m	Maximum values
G	Gas
W	Wall
∞	Ambient, at large distance

Definition of Non-dimensional Numbers

$C_D = F_D/(\rho A U^2)$	Drag coefficient
$C_L = F_L/(\rho A U^2)$	Lift coefficient
$c_M = 4M/(\rho\omega^2 R^5)$	Torque coefficient
$c_w = W/(\rho\, U^2\, L^2)$	Drag coefficient
$\mathrm{Fr} = U^2/(g\,L)$	Froude number
$Re = Ud/\nu$	Reynolds number
$Re_\omega = \rho\omega R^2/\eta$	Rotational Reynolds number
$\mathrm{Pr} = \eta c/k = \nu/a$	Prandtl number
$Pe = \mathrm{Re}\ \mathrm{Pr}$	Peclet number

Chapter 1
Some Fundamentals of Dimensional Analysis

1.1 Some Preliminary Remarks

Why can't a mouse in the size of an elephant stand without breaking its legs? And why can an ant carry multiples of its own weight while humans are barely able to carry more than their own weight? Is this reasoning somehow connected to the fact that the legs of a mouse are "relatively thin" or the weight of the elephant is "relatively large"? One would need to ask: "Relative to what?" Obviously, it appears rather meaningless to claim a length to be "relatively short" or a weight to be "relatively heavy" without defining a suitable scale. But what constitutes an appropriate scale and how may it be found?

In natural and engineering sciences, the common way to measure is by comparison to some agreed standard. For instance, lengths are compared in terms of the unit "meter" and one would say: "This length measures 5 m." This means that the regarded length is five times as long as 1 m. Just as well one might compare the same length with the "millimeter" scale and state: "This length measures 5000 mm." By specifying a numerical value and the applied unit, the considered physical quantity is uniquely described. In practice, units are usually chosen, such that the numerical values can be conveniently used in calculations. However, depending on the applied unit, different values are obtained, to describe exactly the same physical property, in our example the length. Now, on the one hand, physical events are obviously entirely independent of the unit of measure that we arbitrarily have chosen. On the other hand, in order to quantify a physical quantity we are forced to pick some comparative measure.

The way out of this dilemma is to simply compare the regarded length to another length scale that is relevant to the problem of interest. Defining a length through its ratio to some problem related length scale not only provides a unique numerical value, but it also becomes independent of whether one chooses to apply "millimeters" rather than "meters". These considerations are of course also true for all other physical quantities and their respective units. Thus, only the comparison of

V. Simon et al., *Dimensional Analysis for Engineers*,
Mathematical Engineering, DOI 10.1007/978-3-319-52028-5_1

different quantities of the same type leads to meaningful statements and conclusions. This approach implies one important consequence. For illustration let us assume that the regarded problem is governed by two lengths. By relating the two lengths to one another, now instead of two variables, only one, that is the length ratio, describes the problem. In a mathematical sense, therefore, the number of variables is reduced. This turns out to be extremely useful. By reducing the number of characteristic variables the description of the problem is simplified and a solution to the problem may be found more easily. Besides, fewer experiments are required to reveal the complete physical conduct. Now the content of dimensional analysis comprises the precise formulation of this outlined and admittedly vastly simplified approach as well as its scientific reasoning and generalization.

Dimensional analysis holds great benefits, as physical problems can be significantly eased. Application of dimensional analysis removes unessential information from the regarded problem (namely the system of units), thus reducing the number of variables and therefore sharpening the insight to the essential physical mechanisms. In consequence it encourages a deeper understanding and insight into the physical processes. Besides the pure joy of cognition, the engineering task of constitutive and targeted design of technical processes is facilitated.

We want to illustrate this with an example and consider the technically important problem to determine the shaft torque M for a rotating disk immersed in a large body of otherwise quiescent fluid. The system to be studied is shown schematically in Fig. 1.1.

We are interested in the shaft torque M as a function of disk radius R, fluid density ρ, angular velocity ω and fluid viscosity η. Thus, it is sought for the function

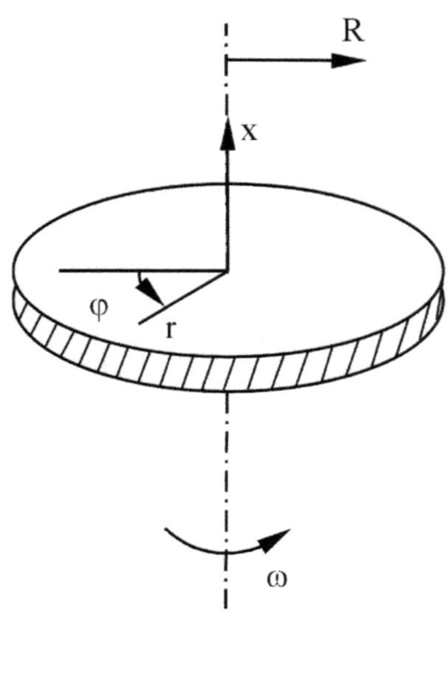

Fig. 1.1 Physical model and coordinate system for the rotating disk

$$M = f(\omega, \eta, \rho, R) \tag{1.1}$$

that determines the functional relation between the shaft torque and the other quantities. So far it has not been possible to obtain the relationship given by Eq. (1.1) for the entire range of industrial interest analytically, so that one has to rely upon experimental measurements.

One might be tempted to determine the value of M for various ω at fixed values of η, ρ and R in order to obtain a relating curve. By additionally varying η, we would obtain a series of curves. For each change of ρ one obtains different sets of curves that could encompass an entire book, and now when one additionally varies R, one would obtain several books filled with plenty of sets of curves. On the contrary, by applying the methods of dimensional analysis the sought relation of Eq. (1.1) can be reduced to

$$c_M = f(\mathrm{Re}_\omega) \tag{1.2}$$

with the two dimensionless quantities

$$c_M = \frac{4M}{\rho\omega^2 R^5} \tag{1.3}$$

and

$$\mathrm{Re}_\omega = \frac{\rho\,\omega\,R^2}{\eta}, \tag{1.4}$$

where c_M denotes the dimensionless shaft torque coefficient and Re_ω is the rotational Reynolds number. Thus, the dimensionless shaft torque can be represented by a single curve that is valid for arbitrary values of ω, η, ρ and R.

Thus, it is sufficient to carry out measurements for a single configuration of one rotational disk and just varying the angular velocity ω. Even though it is not possible to obtain the actual functional relation given by Eq. (1.2) from dimensional analysis the benefit yet becomes apparent: the relation by Eq. (1.1) including 5 variables could be reduced to Eq. (1.2) that now depends on two variables only. The effort to determine the unknown function has thus been considerably simplyfied.

With this simple example, the core message of dimensional analysis is already visible: relations between physical quantities can be traced back to relations between dimensionless variables, or, more precisely: Physical processes are not described by dimensional, but rather by non-dimensional quantities.

The method of dimensional analysis is generally valid. It always provides accurate results. The method is applicable in all branches of physics, whether in mechanics, fluid mechanics, thermodynamics, electricity, optics or other areas. It is useful in chemistry and biology, it is of great value in engineering and is indispensable in model theory and experimentation. In many areas, the systematic application of dimensional analysis has led to great progress, such as in problems of heat transfer and fluid flow engineering. Dimensional considerations were particularly fruitful and fundamentally improved insight in the field of turbulent flows.

Certainly, in the application of dimensional analysis one should not proceed carelessly! The simplicity of the method sometimes leads to believe that its sole use alone already brings physical insight. But that is certainly not the case. It is rather necessary to deeply investigate the problem, to abstract and simplify. Only when we clearly understand the dominant physical mechanisms for a specific problem and insignificant effects are left aside, dimensional analysis provides valuable support, as it allows reducing the number of variables.

From a historical point of view, dimensional analytical approaches date back to the origins of mechanics. Great progress was made in 1686 by Newton and his fundamental work on the motion of bodies (see Görtler 1975). The classical mechanics, founded by him, is based on the three fundamental quantities length, mass and time, which were regarded as fundamental and independent of each other. The notion of physical dimension was only introduced in 1822 by Fourier. It took another 50 years until in 1873 von Helmholtz determined the essential dimensionless products for hydrodynamics in a systematic manner. Ten years later, Reynolds published his studies on the transition of laminar to turbulent flow and related this to the Reynolds number which perhaps represents today's best known dimensionless number (see Görtler 1975). Towards the end of the 19th and beginning of the 20th century the research activity in the field of dimensional analysis tremendously increased. It probably did not enter into general consciousness until 1914 when Buckingham established its general formulation and justification in the essay "On physically similar systems", Buckingham (1914). The master of dimensional analysis is undoubtedly Lord Rayleigh. In his essay "The principles of similitude" published in 1915 he gives numerous examples of the application of dimensional analysis, Rayleigh (1915). Amongst others, on the basis of dimensional considerations he explains why the sky appears to be blue. Since the time after Buckingham's publication, dimensional analysis became a firm pillar in the building of physics. It has been included in textbooks and curricula of natural and engineering sciences. It has—as already mentioned—made significant contributions to the development of many fields of knowledge and is of particular importance in engineering applications.

The present book is divided into four parts. In this first chapter necessary terms are explained, definitions are introduced and the method is derived and explained. This is followed by Chap. 2 on model theory. In the third chapter some examples from different areas are given for illustration. Here one can learn for instance why a mouse in the size of an elephant cannot stand without breaking its legs. Finally, Chap. 4 is concerned with the application of the method to a particular class of differential equations, which lead to similarity solutions. These are particularly important in engineering applications, where dimensional analysis plays an essential role, yet additional considerations are required to a great extent. The appendix contains a collection of exercises that accompany the discussion in the book.

Thus, the overall aim of this booklet is to provide to the reader a practical guide to the fascinating topic of dimensional analysis. Practical guide means that we aim to show “how to do and how to use” the method under consideration.

For those who want to deal with the method of dimensional analysis beyond this practical guide, we recommend to consult the books by Görtler (1975) (with an emphasis on the mathematical theory and proof), Langhaar (1951) (an introduction with many examples) and Spurk (1992) (with a focus and exhaustive treatment of applications in fluid mechanics). For the connoisseurs Lord Rayleigh’s (1915) treatise of 1915 is highly recommended.

1.2 Terms and Definitions

1.2.1 Physical Quantities

Physics and engineering sciences deal with laws that connect different physical quantities. In general terms physical quantities denote some measurable characteristics or properties of physical objects (objects, states or processes). The objects themselves are not necessarily measurable, but their characteristics are. So, for instance, a room per se cannot be measured, yet its height, width, length or volume can be measured. Just as well, a mass point is not measurable by itself, but its coordinates, its velocity or its mass can be quantified. Further, heat conduction is not measurable, but a temperature distribution, the temperature gradient or the thermal conductivity represent measurable quantities. The term measurable quantity also includes variables that cannot be measured directly, but are determined from other measured quantities (Table 1.1).

1.2.2 Entities, Base Quantities, Derived Quantities and Quantity System

The qualitative property of a physical quantity is called its entity. For example, the duration between the maximum deflections of a pendulum is of the entity time. Also the duration in which a cyclist travels a certain distance is of the entity time. The height of a space, the diameter of a sphere, as well as the distance between two points all belong to the same physical entity, which is the length. Physical entities that are independent of each other are called base quantities. The entirety of all base quantities that are necessary to describe the regarded physical relations are referred to as quantity system. Generally, in physics the base quantities for length L, mass M, time T, temperature Θ, electric current I, amount of substance N and the luminous intensity J are used as basic set of quantities. The choice of this $[L\,M\,T\,\Theta\,I\,N\,J]$ system of quantities is by no means mandatory, but rather represents pure

Table 1.1 Some often used physical quantities and their dimensions

Physical quantity	Symbol	Dimension
Velocity of sound	a_s	LT^{-1}
Acceleration	a	LT^{-2}
Angular velocity	ω	T^{-1}
Diffusion coefficient	D	L^2T^{-1}
Density	ρ	ML^{-3}
Dynamic viscosity	η	$ML^{-1}T^{-1}$
Energy	E	ML^2T^2
Frequency	f	T^{-1}
Force	F	MLT^{-2}
Heat transfer coefficient	h	$MT^{-3}\vartheta^{-1}$
Power	P	ML^2T^{-3}
Pressure	p	$ML^{-1}T^{-2}$
Resistance force	W	MLT^{-2}
Surface tension	σ	MT^{-2}
Thermal conductivity	k	$ML^2T^{-3}\vartheta^{-1}$
Thermal diffusivity	α	L^2T^{-1}
Velocity	U	LT^{-1}

convention. The appropriate choice of a quantity system according to the regarded problem is discussed later. Based on these base quantities the so called derived quantities may be deduced, either resulting from physical laws or resulting from useful definitions. Hence, the derived quantities are power law products of the base quantities. As an example, consider the quantity of density, which is the ratio of mass to volume, thus mass divided by the cube of the length. Another example is quantity for velocity, which follows from length divided by time.

1.2.3 Dimension, Dimensional Formula and Bridgman Equation

Following J.C. Maxwell's notation (see Maxwell 1871) the dimension of a physical entity is denoted by square brackets. For example, for the height h of a room, one writes

$$[h] = L. \tag{1.5}$$

The dimension of a derived physical quantity indicates its relation to the base quantities and can be written as a product of powers of the base quantities. The dimension of the velocity u, for instance, follows to

$$[u] = \frac{L}{T} = L^1\,T^{-1} = L\,T^{-1}, \tag{1.6}$$

and one finds that "the dimension of velocity is length times time to the power of minus one". Equation (1.6) is also referred to as a dimension formula. For any derived physical quantity x the dimension formula (1.6) can be written in generalized form as

$$[x] = \prod_{i=1}^{m} X_i^{a_i}, \tag{1.7}$$

where X_i is the ith base quantity of the fundamental quantity system consisting of m base quantities, and a_i is referred to as dimension exponent. Equation (1.7) is also referred to as the Bridgman equation. It is of fundamental importance in dimensional analysis. For a proof of the Bridgeman equation the reader is referred e.g. to Görtler (1975). In particular in the $[L\,M\,T\,\Theta\,I\,N\,J]$ system of quantities, Eq. (1.7), takes the form

$$[x] = L^{a_1}\,M^{a_2}\,T^{a_3}\,\Theta^{a_4}\,I^{a_5}\,N^{a_6}\,J^{a_7}. \tag{1.8}$$

If all dimension exponents a_i of a physical quantity x are zero, Eq. (1.7) results in

$$[x] = \prod_{i=1}^{m} X_i^0 = 1, \tag{1.9}$$

and the variable x is called "dimensionless". Obviously, in particular the ratio of two physical quantities of the same entity must be dimensionless.

1.2.4 Units, Unit System and Unit of Measurement

While the qualitative property of a physical quantity is described by its entity, the quantitative value is determined by a numerical value (measured value) and a unit. Numerical value and unit are thereby correspondingly multiplied. The units of the base quantities are called base units. Today, the $[L\,M\,T\,\Theta\,I\,N\,J]$ system of quantities is generally applied in terms of the corresponding SI-system "Système International d'Unités". This system was established in 1960 internationally and consists of the base units meter (m), kilogram (kg), second (s), Kelvin (K), ampere (A), mole (mol), and candela (cd) (see DIN 1305 from 1974). The unit of a physical quantity is indicated by curly braces around the symbols, so for example

Table 1.2 Base quantities and base unit in the SI-Unit-System

Entity	Symbol	SI-unit	
Length	L	Meter	m
Mass	M	Kilogram	kg
Time	T	Second	s
Temperature	Θ	Kelvin	K
Current	I	Ampere	A
Substance amount	N	Mole	mol
Luminous intensity	J	Candela	cd

$$\{u\} = \frac{m}{s} = m^1\, s^{-1}. \tag{1.10}$$

The entirety of the base units is called unit system, and the entirety of the base quantities and base units is referred to as measuring system. In Table 1.2 the base quantities and base units of the SI-system are summarized.

1.2.5 Change of the Unit System

The use of the SI-unit-system is just as little mandatory as is the choice of the $[L\,M\,T\,\Theta\,I\,N\,J]$-system. Obviously, the distance between two points does not change when measured in meters, in inches or even in Angstroms. However, the numerical value of the physical quantity changes, depending on the used unit system. Denoting a physical quantity with p, its numerical value by x and the base unit by X_i, the variable p can be expressed in the form

$$p = x \prod_{i=1}^{m} X_i^{a_i}. \tag{1.11}$$

Introducing a "new" unit system with new base units X_i' and the corresponding numerical value x', Eq. (1.11) can be rewritten according to

$$p = x' \prod_{i=1}^{m} X_i'^{a_i}. \tag{1.12}$$

The base units in the new and old unit system shall be related by

$$X_i = \alpha_i X_i', \quad (i = 1\ldots m), \tag{1.13}$$

where the scale factors α_i indicate how many units of X_i' are needed to express one unit of X_i. From the identity of the physical quantity p in both systems it follows

$$x \prod_{i=1}^{m} X_i^{a_i} = x' \prod_{i=1}^{m} X_i'^{a_i}, \tag{1.14}$$

and with Eq. (1.13) also

$$x \prod_{i=1}^{m} \alpha_i^{a_i} X_i'^{a_i} = x' \prod_{i=1}^{m} X_i'^{a_i}, \tag{1.15}$$

so that the relation between the numerical values in the old and the new system becomes

$$x' = x \prod_{i=1}^{m} \alpha_i^{a_i}. \tag{1.16}$$

As an example we consider the velocity of a bike, when one drives fast down the hill U, whose dimension is given in the $[LMT\Theta INJ]$ system by $[U] = LT^{-1}$. Let us initially measure U in the units m and s, and in the new system in km and h, so that we obtain

$$U = \widehat{U}\,\mathrm{m\,s^{-1}} \quad \text{and} \quad U = \widehat{U}'\,\mathrm{km\,h^{-1}} \tag{1.17}$$

with a value for our bike ride of $\widehat{U} = 15.0$, and we seek the measure $\hat{U}'$ in the new system. For this example Eq. (1.13) reads

$$\mathrm{m} = \frac{1}{1000}\,\mathrm{km} \quad \text{and} \quad \mathrm{s} = \frac{1}{3600}\,\mathrm{h}, \tag{1.18}$$

from which the scale factors

$$\alpha_1 = \frac{1}{1000} \quad \text{and} \quad \alpha_2 = \frac{1}{3600} \tag{1.19}$$

are obtained. The numerical value $\hat{U}'$ follows from Eq. (1.16) to

$$\widehat{U}' = \widehat{U}\,\alpha_1^1\,\alpha_2^{-1} = 15.0\frac{1}{1000}\left(\frac{1}{3600}\right)^{-1} = 54.0. \tag{1.20}$$

With the prior considerations, only those features of a physical quantity were considered that arise from the demand for the measurability of these quantities. But physical quantities also have other properties, such as vector or tensor properties. These properties cannot be included by comparison with a scale, and assigning a respective numerical value. Although one may try to consider vector properties for the purposes of dimensional analysis by introducing three basic units of length, and under certain circumstances it allows sharpening the result (see Langhaar 1951). In general, this is however not possible, so that vector and tensor properties remain disregarded in dimensional analytical considerations.

1.2.6 Physical Equations and Dimensional Constants

A very fundamental assumption in physics is that physical principals can be expressed in terms of mathematical equations that describe the underlying relation of different physical quantities. Mathematical equations that reflect physical laws are referred to as physical equations. For physical equations to remain valid for any chosen unit system, each summand of such an equation must carry the same dimension. Equations which satisfy this property are called dimensionally homogeneous. When changing the unit system in a physical equation, according to Eq. (1.13) and introducing a unit changed by a factor α, then the numerical values of all summands are correspondingly changed by a constant factor. Then this constant factor can be cancelled in all terms, and the equation maintains its original form (see also Langhaar 1951).

In the context of physical considerations, the addition (or subtraction) of different entities is not allowed, because it does not lead to physical equations. Such equations are not dimensionally homogeneous since an alternate choice of units does not necessarily change the numerical values of all summands by the same factor.

In engineering sciences sometimes empirical equations or numerical value equations are used. When repeatedly applying the same equation, that contains constants and material values, it is often perceived as practical to calculate the recurring numerical values. As an example, we use the thermal state equation of an ideal gas (see e.g. Weigand et al. 2016)

$$pv = RT \tag{1.21}$$

that determines the relation between pressure p, specific volume v and temperature T of an ideal gas. In this equation, we have also the gas constant R, which takes different values for each gas. Equation (1.21) is dimensionally homogeneous, and therefore valid for any choice of unit system. However, if we are introducing the numerical value of R = 287 J/(kg K) for air, we obtain from Eq. (1.21)

$$pv = 287\,T, \quad \text{where}\ \{p\} = \mathrm{kg\,m^{-1}s^{-2}}, \{v\} = \mathrm{kg^{-1}\,m^3}, \text{and}\ \{T\} = \mathrm{K}. \tag{1.22}$$

This is a numerical value equation, which is valid only for the units meter, second and Kelvin. Equation (1.22) may be regarded as an empirically derived equation, where the experimenter has measured the ratio of p, v and T and found that this ratio is constant for air. Although this equation can be quite useful for practical purposes, it is not dimensionally homogeneous, and therefore does not constitute a physical equation. Equation (1.22) may be converted back into a dimensionally homogeneous form by interpreting the parameter 287 in Eq. (1.22) as a dimensional quantity with a numerical value of 1 and the unit $m^2s^{-2}K^{-1}$.

This consideration can be generalized: any empirical equation can be converted into a dimensionally homogeneous form by introducing appropriate dimensional constants. In fact, the laws of physics are dimensionally homogeneous because they are brought in a dimensionally homogeneous form by introducing corresponding dimensional constants.

Variables that change their value with alternate choice of base units are to be counted as additional variables in dimensional analytical considerations. Therefore, in addition to the variables in a strict mathematical sense one must also include material values and dimensional constants.

Therefore, for dimensional analytical considerations the following is important: The methods of dimensional analysis can be only applied to dimensionally homogeneous equations, and dimensional constants (for instance, to bring dimensionally non-homogeneous equations in a dimensionally homogeneous form) are to be counted as variables in the sense of dimensional analysis.

1.3 Buckingham-Theorem (Pi-Theorem)

The preceding considerations can be summarized as follows: Physical processes, systems or objects are described by physical quantities. The numerical values of the physical quantities, however, change when varying the units in which they are measured. But the physical systems cannot depend on the choice of units in which the physical quantities are measured. The physical systems can therefore only depend on dimensionless ratios of physical quantities of the same entity, as their numerical values remain unchanged upon a change of units. These dimensionless quantities are denoted in the following by $\prod$ (Spurk 1992; Buckingham 1914).

This statement can also be formulated mathematically (in the following we follow the derivation given by Spurk (1992) since it is very clear and focused): We want to determine a physical quantity p_1 as a function of the physical variables $p_2, p_3, \ldots, p_n$, where n is the number of physical quantities describing the problem. The assignment therefore reduces to a relationship of the form

$$p_1 = f(p_2, p_3, \ldots, p_n), \tag{1.23}$$

respectively

$$f(p_1, p_2, p_3, \ldots, p_n) = 0. \tag{1.24}$$

Since a physical process or condition must be independent of the chosen units, it can be reduced to a relationship between dimensionless quantities $\prod_i$ such that Eq. (1.24) takes the equivalent form

$$f\left(\prod_1, \prod_2, \prod_3, \ldots, \prod_d\right) = 0, \tag{1.25}$$

where d is the number of dimensionless products. We shall see later in this chapter, how d is related to n and how it can be determined. The equivalence of Eqs. (1.24) and (1.25) is referred to as "Buckingham theorem" or as "Pi-Theorem" (Spurk 1992; Buckingham 1914).

The equivalence of Eqs. (1.24) and (1.25) is shown in the following. The dimensionless products are products of powers of the physical quantities p_j

$$\prod = \prod_{j=1}^{n} p_j^{k_j} \tag{1.26}$$

with the dimension

$$\left[\prod\right] = 1 = \prod_{j=1}^{n} [p_j]^{k_j}. \tag{1.27}$$

Simultaneously, for the dimension of the physical quantities p_j it follows that

$$[p_j] = \prod_{i=1}^{m} X_i^{a_{ij}}, \quad (j = 1\ldots n), \tag{1.28}$$

such that Eq. (1.27) takes the form

$$\left[\prod\right] = 1 = \prod_{j=1}^{n} \prod_{i=1}^{m} X_i^{a_{ij}\,k_j}. \tag{1.29}$$

The coefficients a_{ij} are the exponents of the i-th base quantity in the dimension formula for the j-th physical quantity and thus these coefficients are known.

For known dimension exponents a_{ij}, the task is to determine the exponents k_j such that Eq. (1.29) is satisfied. Equation (1.29) is exactly satisfied when the sum of the exponents of the i-th base quantity is zero, i.e. when the system of linear equations

$$\sum_{j=1}^{n} a_{ij}\,k_j = 0, \quad (i = 1\ldots m) \tag{1.30}$$

has non-trivial solutions for the n unknowns k_j. Written in matrix form (with row index i and column index j) the coefficient matrix (a_{ij}) becomes

$$(a_{ij}) = \begin{bmatrix} a_{11} & a_{12} & \cdots & a_{1n} \\ a_{21} & a_{22} & & a_{2n} \\ \vdots & & & \vdots \\ a_{m1} & \cdots & & a_{mn} \end{bmatrix} \tag{1.31}$$

The coefficient matrix (a_{ij}) is also referred to as **dimension matrix**. A square matrix has a determinant

$$D = \det(a_{ij}). \tag{1.32}$$

If the determinant vanishes the matrix is called singular, otherwise it is called regular, e.g. Bronstein and Semendjajew (1981). Generally dimension matrices are of rectangular shape $(n > m)$. In that case, by crossing out rows and columns of the initial matrix (a_{ij}), a regular square submatrix is formed that consists of r linearly independent rows and columns, so that the determinant does not vanish.

If all square sub-matrices with line numbers larger than r are singular, the dimension matrix has the rank r and the defect $d = n - r$. A linear homogeneous system of equations for n unknowns, whose coefficient matrix has the rank r consists of r linearly independent equations and has $d = n - r$ linearly independent non-trivial solutions. Thus, there are more unknowns to be determined than independent equations are available. Thus, one can arbitrarily choose the $d = n - r$ unknowns and from the homogeneous system

$$\sum_{j=1}^{n} a_{ij}\, k_j = 0, \quad (i = 1\ldots r) \tag{1.33}$$

determine the remaining r unknowns. For that purpose Eq. (1.33) is written in terms of the inhomogeneous system of linear equations

$$\sum_{j=n-r+1}^{n} a_{ij} k_j = -\sum_{j=1}^{n-r} a_{ij} k_j = b_i, \quad (i = 1\ldots r) \tag{1.34}$$

with r unknowns. On the right hand side b_i is known by assigning arbitrary values for k_j, $(j = 1\ldots n - r)$. The system of Eq. (1.34) has a unique solution, e.g. using Cramer's rule (Bronstein and Semendjajew 1981)

$$k_l = D_l / D, \tag{1.35}$$

where the determinant

$$D = \det(a_{ij}) \neq 0, \quad (i = 1\ldots r, \quad j = 1\ldots r) \tag{1.36}$$

and D_l is the determinant of the matrix obtained by replacing the $D = \det(a_{ij})$-th column of the coefficient matrix by the right side b_i. The r solutions of Eq. (1.34) together with the $d = n - r$ freely selected unknowns now represent the n-dimensional solution vector $k_{(i),j}, (i = 1\ldots d; \quad j = 1\ldots n)$, that follows when arbitrarily assigning the d unknowns respectively by d linearly independent values, that thus determine the r unknowns in Eq. (1.34). Linear independence is ensured when the matrix of the selected unknowns is regular, that is, when

$$\det \begin{bmatrix} k_{(1),1} & k_{(1),2} & \cdots & k_{(1),d} \\ k_{(2),1} & k_{(2),2} & & k_{(2),d} \\ \vdots & & & \vdots \\ k_{(d),1} & \cdots & & k_{(d),d} \end{bmatrix} \neq 0. \tag{1.37}$$

For a systematic calculation it is suitable to allocate the unknowns in terms of the unit matrix, thus

$$\begin{bmatrix} k_{(1),1} & k_{(1),2} & \cdots & k_{(1),d} \\ k_{(2),1} & k_{(2),2} & & k_{(2),d} \\ \vdots & & & \vdots \\ k_{(d),1} & \cdots & & k_{(d),d} \end{bmatrix} = \begin{bmatrix} 1 & 0 & \cdots & 0 \\ 0 & 1 & & 0 \\ \vdots & & & \vdots \\ 0 & 0 & \cdots & 1 \end{bmatrix}. \tag{1.38}$$

For the dimensionless products with Eq. (1.26) we then obtain

$$\begin{aligned} \prod\nolimits_1 &= p_1^1 p_2^0 p_3^0 \quad \cdots \quad p_d^0 p_{d+1}^{k_{(1),d+1}} \quad \cdots \quad p_n^{k_{(1),n}}, \\ \prod\nolimits_2 &= p_1^0 p_2^1 p_3^0 \quad \cdots \quad p_d^0 p_{d+1}^{k_{(2),d+1}} \quad \cdots \quad p_n^{k_{(2),n}}, \\ &\vdots \qquad\qquad\qquad\quad \vdots \\ &\text{and} \\ \prod\nolimits_d &= p_1^0 p_2^0 p_3^0 \quad \cdots \quad p_d^1 p_{d+1}^{k_{(d),d+1}} \quad \cdots \quad p_n^{k_{(d),n}}. \end{aligned} \tag{1.39}$$

We now consider these expressions to be solved with regard to p_1, p_2 to p_d and after introducing each one in Eq. (1.25) we obtain

$$f\left(\prod\nolimits_1 \cdots \prod\nolimits_d, p_{d+1} \cdots p_n\right) = 0. \tag{1.40}$$

Now we change the units of the base quantities. Then the numerical values of the physical quantities change according to

$$p'_j = p_j \prod_{i=1}^{r} c_i^{a_{ij}}, \quad (j = 1\ldots n) \tag{1.41}$$

and Eq. (1.40) reads

$$f\left(\prod_1 \cdots \prod_d, p'_{d+1} \prod_{i=1}^{r} c_i^{-a_{i,d+1}} \ldots p'_n \prod_{i=1}^{r} c_i^{-a_{i,n}}\right) = 0. \tag{1.42}$$

The products

$$g_j = p'_j \prod_{i=1}^{r} c_i^{-a_{ij}}, \quad (j = d+1\ldots n) \tag{1.43}$$

can be assigned to any numerical value, since the system of equations

$$\sum_{i=1}^{r} a_{ij} \ln(c_i) = \ln(p'_j) - \ln(g_j), \quad (j = d+1\ldots n) \tag{1.44}$$

has a unique solution. The coefficient matrix of Eq. (1.44) is the transpose of the regular part of the matrix of rank r of system (1.34), and therefore is regular itself, and the right side of Eq. (1.44) is not zero for all j. Solving Eq. (1.44) yields a unique solution. But if any value can be assigned to the products g_j the function according to Eq. (1.42) cannot depend on them, so that from Eq. (1.42) finally we obtain

$$f\left(\prod_1 \cdots \prod_d\right) = 0. \tag{1.45}$$

It is important to note here that the solution is not unique, since the assignment by Eq. (1.38) is not unique. The d solution vectors form a fundamental system, i.e. any solution of the system (1.34) can be represented by the linear combination

$$k'_{(m),j} = \sum_{i=1}^{d} C_{(m),i} k_{(i),j}, \quad (j = 1\ldots n) \tag{1.46}$$

with regular $C_{(m),i}$. Thus, there exist an infinite number of solutions, each of which consists of d linearly independent vectors. To illustrate the calculation process we exemplarily return to the problem of the rotating disk. The shaft torque M is sought for the rotating disk as a function of the flow angular velocity ω, the viscosity of the liquid η, the density of the liquid ρ and of the disk radius R, hence

$$M = f(\omega, \eta, \rho, R). \tag{1.47}$$

In the $[LMT]$-system, one obtains the dimension matrix

	M	ω	η	ρ	R
	k_1	k_2	k_3	k_4	k_5
L	2	0	−1	−3	1
M	1	0	1	1	0
T	−2	−1	−1	0	0

In this matrix, the quantities that are to occur linearly in the dimensionless products were placed to the left. The rank of the dimension matrix is $r = 3$, so that $d = n - r = 2$ dimensionless products are found. For the exponents of the basic variables to become zero, the relation

$$\sum_{j=1}^{n} a_{ij} k_j = 0 \tag{1.48}$$

must be satisfied. In the present case this yields the following system of equations

$$\begin{bmatrix} 2 & 0 & -1 & -3 & 1 \\ 1 & 0 & 1 & 1 & 0 \\ -2 & -1 & -1 & 0 & 0 \end{bmatrix} \begin{bmatrix} k_1 \\ k_2 \\ k_3 \\ k_4 \\ k_5 \end{bmatrix} = 0 \tag{1.49}$$

and from this

$$\begin{bmatrix} 2 & 0 \\ 1 & 0 \\ -2 & -1 \end{bmatrix} \begin{bmatrix} k_1 \\ k_2 \end{bmatrix} + \begin{bmatrix} -1 & -3 & 1 \\ 1 & 1 & 0 \\ -1 & 0 & 0 \end{bmatrix} \begin{bmatrix} k_3 \\ k_4 \\ k_5 \end{bmatrix} = 0 \tag{1.50}$$

respectively

$$\begin{bmatrix} -1 & -3 & 1 \\ 1 & 1 & 0 \\ -1 & 0 & 0 \end{bmatrix} \begin{bmatrix} k_3 \\ k_4 \\ k_5 \end{bmatrix} = - \begin{bmatrix} 2 & 0 \\ 1 & 0 \\ -2 & -1 \end{bmatrix} \begin{bmatrix} k_1 \\ k_2 \end{bmatrix} \tag{1.51}$$

and after conversion finally

$$\begin{bmatrix} k_3 \\ k_4 \\ k_5 \end{bmatrix} = -\begin{bmatrix} 0 & 0 & -1 \\ 0 & 1 & 1 \\ 1 & 3 & 2 \end{bmatrix} \begin{bmatrix} 2 & 0 \\ 1 & 0 \\ -2 & -1 \end{bmatrix} \begin{bmatrix} k_1 \\ k_2 \end{bmatrix} \tag{1.52}$$

thus

$$\begin{bmatrix} k_3 \\ k_4 \\ k_5 \end{bmatrix} = \begin{bmatrix} -2 & -1 \\ 1 & 1 \\ -1 & 2 \end{bmatrix} \begin{bmatrix} k_1 \\ k_2 \end{bmatrix}. \tag{1.53}$$

Now a set of $d = n - r = 2$ linearly independent solution vectors $k_{(i),j}$, $(i = 1\ldots d, j = 1\ldots n)$ is obtained by assigning the $k_{(i),j}$, $(i, j = 1\ldots d)$ to unit vectors on the right hand side of Eq. (1.53) according to

$$\begin{bmatrix} k_{(1),1} & k_{(1),2} \\ k_{(2),1} & k_{(2),2} \end{bmatrix} = \begin{bmatrix} 1 & 0 \\ 0 & 1 \end{bmatrix}. \tag{1.54}$$

By substituting Eq. (1.54) into Eq. (1.53), the $k_{(i),1}, k_{(i),2}, k_{(i),3}$, $(i = d + 1\ldots n)$ can be determined, as summarized in the following table

	k_1	k_2	k_3	k_4	k_5
Π_1	1	0	−2	1	−1
Π_2	0	1	−1	1	2

The dimensionless products then follow from

$$\prod_i = M^{k_{(i),1}} \omega^{k_{(i),2}} \eta^{k_{(i),3}} \rho^{k_{(i),4}} R^{k_{(i),6}}, \quad (i = 1\ldots d) \tag{1.55}$$

to

$$\prod_1 = \frac{M\rho}{\eta^2 R} \quad \text{and} \quad \prod_2 = \frac{\omega R^2}{\nu}, \tag{1.56}$$

where the kinematic viscosity is given by

$$\nu = \frac{\eta}{\rho}. \tag{1.57}$$

Equation (1.47) can therefore now be written in the form

$$\frac{M\,\rho}{\eta^2 R} = f\left(\frac{\omega\, R^2}{\nu}\right). \tag{1.58}$$

The dimensionless product $\prod_1$ is, however, not used in practice. Since the two solution vectors $k_{(i),j}$, $(i = 1\ldots2;\ j = 1\ldots5)$ represent a fundamental system, according to Eq. (1.46) the new solution

$$k'_{(1),j} = k_{(1),j} - 2k_{(2),j} \tag{1.59}$$

can be obtained and using Eq. (1.26), the new dimensionless products

$$\prod\nolimits'_1 = \prod_{j=1}^{5} p_j^{k_{(1),j} - 2\,k_{(2),j}} = \prod_{j=1}^{5} p_j^{k_{(1),j}} \prod_{j=1}^{5} p_j^{-2k_{(2),j}} = \prod\nolimits_1 \prod\nolimits_2^{-2}, \tag{1.60}$$

or

$$\prod\nolimits'_1 = \frac{M}{\rho \omega^2 R^5}. \tag{1.61}$$

Thus, Eq. (1.58) yields the relation

$$\frac{M}{\rho\, \omega^2\, R^5} = f\left(\frac{\omega R^2}{\nu}\right) \tag{1.62}$$

and with Eqs. (1.3) and (1.4) one finally obtains

$$c_M = f(\mathrm{Re}_\omega). \tag{1.63}$$

1.4 Choice of the Base System

In the previous considerations we have always used the $[LMT]$ system. This choice is however not mandatory for the purposes of dimensional analysis. It is only required that the basic parameters of the system are independent in the previously described sense. We have seen in the last section that the number of dimensionless products is given by $d = n - r$, and one might expect that the rank r of the dimensional matrix would increase when increasing the number of base quantities. As a result, the expected number of dimensionless products d would reduce and the problem might be accordingly simplified. However, it turns out that introducing additional numbers of base quantities also increases the number of variables n, so that n and r are simultaneously increased, and the net profit therefore remains the same. This shall be explained in the following by examples.

1.4.1 The [LMFT]-System

Sometimes it can be advantageous to use a $[LMFT]$-system, in which the length, mass, force and time represent basic variables. In the $[LMT]$-system Newton's Second Law applies in the form

$$F = ma, \tag{1.64}$$

in which the force F is required to accelerate a body of mass m by the acceleration a. In order for Newton's Second Law to remain a dimensionally homogeneous equation in the $[LMFT]$-system a dimensionless constant has to be introduced in the form

$$F = C\,ma \tag{1.65}$$

with

$$[C] = L^{-1}M^{-1}F^{1}T^{2}. \tag{1.66}$$

As an example we reconsider the problem of the rotating disk. In the $[LMFT]$-system, the dimensional constant C must be included as a variable, and Eq. (1.47) becomes

$$M = f(\omega, \eta, \rho, R, C), \tag{1.67}$$

so that the dimension matrix is determined to

	M	ω	η	ρ	R	C
L	1	0	−2	−3	1	−1
M	0	0	0	1	0	−1
F	1	0	1	0	0	1
T	0	−1	1	0	0	2

After transformation, one obtains

	$M/(\eta\omega R^3)$	ω	η	ρ	R	C
L	0	1	−2	−3	1	−1
M	0	0	0	1	0	−1
F	0	0	1	0	0	1
T	0	−1	1	0	0	2

respectively

	$M/(\eta\,\omega\,R^3)$	$C\;\omega R/\eta$	η	ρ	R	C
L	0	2	−2	−3	1	−1
M	0	−1	0	1	0	−1
F	0	0	1	0	0	1
T	0	0	1	0	0	2

and eventually

	$M/(\eta\,\omega\,R^3)$	$C\;\rho\omega R^2/\eta$	η	ρ	R	C
L	0	0	−2	−3	1	−1
M	0	0	0	1	0	−1
F	0	0	1	0	0	1
T	0	0	1	0	0	2

Thus, there are two dimensionless products obtained to

$$\prod_1 = \frac{M}{\eta\omega R^3} \quad \text{and} \quad \prod_2 = C\frac{\rho\omega R^2}{\eta}. \tag{1.68}$$

Thus, one obtains

$$\frac{M}{\eta\omega R^3} = f\left(C\frac{\rho\omega R^2}{\eta}\right). \tag{1.69}$$

To return to the $[LMT]$-system the numerical value of the constant C is assigned to unity with the dimension of 1. Equations (1.62) and (1.69) are dimensionally equivalent to each other, because the dimensionless quantities given by Eq. (1.68) can obviously easily be combined to obtain the quantities given in Eqs. (1.56) and (1.61). Thus, again two dimensionless products are obtained, and increasing the number of basic quantities has brought no progress. This is not surprising because by the transformation from the $[LMT]$- to the $[LMFT]$-system, the number of variables increased from $n = 5$ to $n = 6$ and the rank of the dimensional matrix increased from $r = 3$ to $r = 4$ so that in both cases $d = n - r = 2$ dimensionless quantities are to be expected.

In case that the additional dimensional constant C is not introduced as a variable in Eq. (1.67), the dimension matrix reads

	M	ω	η	ρ	R
L	1	0	−2	−3	1
M	0	0	0	1	0
F	1	0	1	0	0
T	0	−1	1	0	0

Now $n = 5$ and $r = 4$, and one obtains only one dimensionless product

$$\Pi_1 = \frac{M}{\eta\omega R^2} = \text{const.} \tag{1.70}$$

The fact that the constant C is not included as a variable implies that Newton's Second Law in the considered problem is irrelevant, so that an acceleration free flow is considered. However that is approximately applicable only for very small rotational Reynolds numbers in the case when the friction forces are much larger than the inertial forces. For larger rotational Reynolds numbers the acceleration forces are no longer negligible, and the flow is accelerated so that Newton's Second Law in dimensionally homogeneous form has to be taken into account, and the constant C has to be included. Thus, Eq. (1.70) might be useful as a solution for very small rotational Reynolds numbers.

1.4.2 The [LMTΘ]-System

Next, as an example the isothermal and incompressible steady flow around a body, with a temperature difference $\Delta\vartheta$ between body and fluid is considered. We are interested in the heat flow $\dot{Q}$ which is transferred from the body to the liquid. The thermal conductivity of the liquid is k, its density is ρ and the specific heat capacity is c. For incompressible flow, however, the density occurs only as a product with the heat capacity. So it is sought for a function of the form

$$\dot{Q} = f\,(\Delta\vartheta, U, k, \rho\, c, d) \tag{1.71}$$

The dimension matrix then reads

	$\dot{Q}$	$\Delta\vartheta$	U	k	ρc	d
L	2	0	1	1	−1	1
M	1	0	0	1	1	0
T	−3	0	−1	−3	−2	0
Θ	0	1	0	−1	−1	0

and we obtain by rearranging the dimension matrix

	$\frac{\dot{Q}}{k\Delta\vartheta d}$	$\Delta\vartheta$	U	k	$\frac{\rho c U d}{k}$	d
L	0	0	1	1	0	1
M	0	0	0	1	0	0
T	0	0	−1	−3	0	0
Θ	0	1	0	−1	0	0

from which the two dimensionless products

$$\prod_1 = \frac{\dot{Q}}{k\,\Delta\vartheta\,d} \tag{1.72}$$

and

$$\prod_2 = \frac{\rho\,c\,U\,d}{k} \tag{1.73}$$

are determined. Thus, we obtain the following relation for the heat flux

$$\frac{\dot{Q}}{k\Delta\vartheta d} = f\left(\frac{\rho\,c\,Ud}{k}\right). \tag{1.74}$$

Now, according to the kinetic theory of gases, the temperature is proportional to the mass m of the molecules and the square of its fluctuation speed v according to

$$\left\langle \frac{m\,v^2}{2} \right\rangle = k_B \vartheta, \tag{1.75}$$

where k_B is the Boltzmann constant (see e.g. Vincenti and Kruger 1986). Thus, one can understand the temperature as a derived quantity with the dimension $[\vartheta] = L^2M^1T^{-2}$. The corresponding dimension of the matrix is gained by substituting the dimension of the temperature $[\vartheta] = L^2M^1T^{-2}$ in all dimension formulas. One therefore obtains

	$\dot{Q}$	$\Delta\vartheta$	U	k	ρc	d
L	2	2	1	−1	−3	1
M	1	1	0	0	0	0
T	−3	−2	−1	−1	0	0

and after transformation

	$\frac{\dot{Q}}{k\Delta\vartheta d}$	$\Delta\vartheta$	$\frac{U}{kd^2}$	k	$\rho\, cd^3$	d
L	0	2	0	−1	0	1
M	0	1	0	0	0	0
T	0	−2	0	−1	0	0

and we identify the three dimensionless products

$$\Pi_1 = \frac{\dot{Q}}{k\,\Delta\vartheta\, d}, \tag{1.76}$$

$$\Pi_2 = \frac{\rho\, c\, U\, d}{k}, \tag{1.77}$$

and

$$\Pi_3 = \rho\, c\, d^3. \tag{1.78}$$

We obtain the relation

$$\frac{\dot{Q}}{k\Delta\vartheta\, d} = f\left(\rho\, c\, d^3, \frac{\rho\, c\, U\, d}{k}\right). \tag{1.79}$$

At first this result seems to contradict Eq. (1.74). This conflict shall now be clarified. In the consideration in the $[LMT]$-system, the temperature was regarded as a derived quantity. But then the constant k_B in Eq. (1.75) is dimensionless. We repeat the entire procedure, but now in the $[LMT\Theta]$-system. For Eq. (1.75) to be a dimensionally homogeneous equation, the dimensional constant k_B must have the dimension $[k] = L^2M^1T^{-2}\Theta^{-1}$ and it must be introduced as an additional variable, so that Eq. (1.71) takes the form

$$\dot{Q} = f(\Delta\vartheta, U, k, \rho\, c, d, k_B) \tag{1.80}$$

The dimension matrix is

	$\dot{Q}$	$\Delta\vartheta$	U	k	$\rho\, c$	d	k_B
L	2	0	1	1	−1	1	2
M	1	0	0	1	1	0	1
T	−3	0	−1	−3	−2	0	−2
Θ	0	1	0	−1	−1	0	−1

Just as in the $[LMT]$ system one obtains three dimensionless quantities with

$$\prod_1 = \frac{\dot{Q}}{k\,\Delta\vartheta\,d}, \tag{1.81}$$

$$\prod_2 = \frac{\rho\,c\,U\,d}{k} \tag{1.82}$$

and

$$\prod_3 = \frac{\rho\,c\,d^3}{k_B}, \tag{1.83}$$

so that Eq. (1.80) can now be written in the form of

$$\frac{\dot{Q}}{k\,\Delta\vartheta\,d} = f\left(\frac{\rho\,c\,d^3}{k_B}, \frac{\rho\,c\,U\,d}{k}\right) \tag{1.84}$$

The apparent contradiction between Eqs. (1.74) and (1.79) can now be resolved: The fact that the dimensional constant k_B is not included in the variable list of Eq. (1.71), implies that equation Eq. (1.75) is considered to be insignificant in the regarded problem. The constant represents the so called Boltzmann's constant, and its numerical value is so small, that in fact in most practical applications the numerical value of $\prod_3$ is so large, that Eq. (1.84) can no longer depend on $\prod_3$ and again equation Eq. (1.74) results. However, when regarding the flow in very dilute gases, Eq. (1.75) may have a significant role, and $\prod_3$ can no longer be neglected.

Finally, the heat transfer problem shall be extended to the case when friction effects are not negligible the fluid is subject to friction and we introduce the viscosity η as an additional variable. Then the additional dimensionless product

$$\prod_4 = \frac{\rho\,c\,\eta}{k\,\rho} = \frac{\rho\,c\,\nu}{k} = \mathrm{Pr} \tag{1.85}$$

is obtained, representing the so called Prandtl number. In practical applications besides $\prod_4$, instead of $\prod_2$ one applies the product

$$\prod_2' = \prod_2 \prod_4^{-1} = \frac{U\,d}{\nu} = \mathrm{Re} \tag{1.86}$$

so that the relation eventually takes the form

$$\frac{\dot{Q}}{k\,\Delta\vartheta\,d} = f(\mathrm{Re}, \mathrm{Pr}). \tag{1.87}$$

Equation (1.87) shows that the dimensional heat flux $\dot{Q}/(k\,\Delta\vartheta\,d)$ is a function of the two dimensional quantities Re and Pr for a forced convective flow. This equation is of prime importance in many heat transfer predictions and the reader is referred to Kays et al. (2004) for a detailed discussion on how the function $f\,$(Re, Pr) can be determined for different heat transfer applications.

Chapter 2
Model Theory

This chapter explains the fundamentals of model theory which can be used for obtaining scaled experiments in a laboratory. In order to obtain complete similarity, all dimensionless products have to be the same in the model experiment and in the technical application. Then results from the model experiment can be transferred immediately to the technical application. If not all dimensionless products can be kept the same in the model experiments compared to the real application incomplete similarity will be obtained. Also here, the results from the model experiments are very useful. This is also shown and discussed in this chapter.

2.1 General Model Theory

The principles of scientific thinking and working involve the observation of the surrounding nature and consequently to develop theories that explain the observed events and predict future behavior. Every theory requires its verification by experiments. The necessary experiments are performed in many applications in physics and the engineering sciences on a basis of scaled models. The spatial extents of the model are often larger or smaller than that of the original case. This is shown in Fig. 2.1.

In many applications, the size of the original object is so large that only scaled models are suitable for experimental investigation. Consider, for example the study of the flow around an aircraft, the flow around skyscrapers or the flow through rivers and lakes. Models with dimensions that are larger than that of the original are displayed if the spatial resolution of the original is too low or the experiment itself would be technically difficult to measure, as for instance in the case of internal cooling of gas turbine blades.

Each model experiment must be preceded by a thorough dimensional analytical consideration of the problem, because the verified physical laws in the experiment depend only on dimensionless quantities. The basics of model theory can be very

V. Simon et al., *Dimensional Analysis for Engineers*,
Mathematical Engineering, DOI 10.1007/978-3-319-52028-5_2

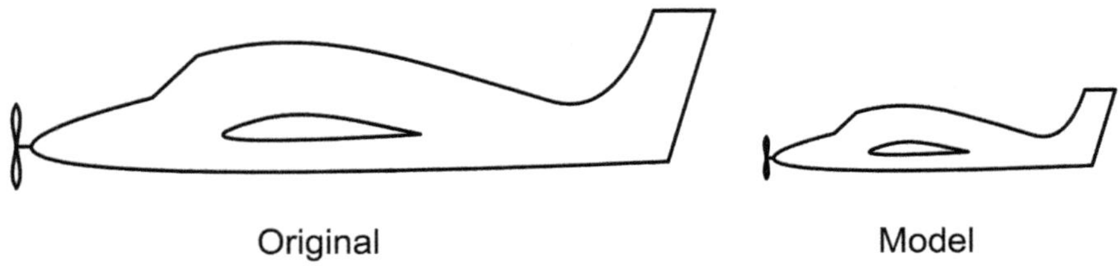

Fig. 2.1 Schematic comparison between geometries for the real application and laboratory testing

easily deduced from the previous considerations. We assume Eq. (1.45) to be solved for one of the dimensionless products (here we follow the derivation of Spurk (1992), since it is very focused):

$$\prod_1 = f\left(\prod_2 \cdots \prod_d\right). \tag{2.1}$$

Obviously $\prod_1$ does not change even when the physical quantities contained in $\prod_i (i = 2 \ldots d)$ change, as long as the dimensionless products remain unchanged. Denoting the physical parameters of the original by p_j and those of the model by p'_j, the relationship between the two follows to

$$p'_j = M_j p_j, \quad (j = 1 \ldots n), \tag{2.2}$$

where M_j denotes the scale factor of the j-th physical quantity. In order to ensure complete physical similarity between model and original, all dimensionless products for model and original must be equal, i.e.

$$\prod_i = \prod'_i, \quad (i = 1 \ldots d). \tag{2.3}$$

With Eq. (2.2) it follows

$$\prod_{j=1}^{n} p_j^{k_{(i),j}} = \prod_{j=1}^{n} p_j^{\prime k_{(i),j}} = \prod_{j=1}^{n} p_j^{k_{(i),j}} \prod_{j=1}^{n} M_j^{k_{(i),j}}, \quad (i = 1 \ldots d), \tag{2.4}$$

so that the scale factors are obtained by

$$\prod_{j=1}^{n} M_j^{k_{(i),j}} = 1, \quad (i = 1 \ldots d). \tag{2.5}$$

By taking the logarithm of this equation it follows

$$\sum_{j=1}^{n} k_{(i),j} \ln\left(M_j\right) = 0, \quad (i = 1\ldots d). \tag{2.6}$$

Equation (2.6) is a system of linear equations with $d = n - r$ equations for the unknown scale factors. According to the considerations that were made in connection with Eq. (1.33), these d equations are linearly independent. One can therefore define $n - d = r$ scale factors arbitrarily, and the remaining scale factors can then be uniquely calculated from Eq. (2.6).

2.2 Complete Similarity

One speaks of "complete similarity" or "full similarity" when all dimensionless products are the same in model and original. As a simple example of complete similarity, we consider the experimental investigation of a wind wheel in the model. The power P of the wind turbine is a function of blade length L, the wind speed U, the rotational speed n and the density of the air ρ, thus

$$P = f\,(L,\, U,\, n,\, \rho). \tag{2.7}$$

The corresponding dimension exponents are summarized in the following dimension matrix

	P	L	U	n	ρ
L	2	1	1	0	−3
M	1	0	0	0	1
T	−3	0	−1	−1	0

This yields two dimensionless products so that Eq. (2.7) takes the form

$$\frac{P}{\rho\, U^3\, L^2} = f\left(\frac{U}{n\, L}\right). \tag{2.8}$$

For the model and the full-scale original to show the same physical behavior, the dimensionless numbers must be equal, i.e. it must be insured that

$$\frac{P}{\rho\, U^3\, L^2} = \frac{P'}{\rho'\, U'^3\, L'^2} \quad \text{and} \quad \frac{U}{n\, L} = \frac{U'}{n'\, L'} \tag{2.9}$$

or

$$\frac{P'}{P} = \frac{\rho'}{\rho}\left(\frac{U'}{U}\right)^3\left(\frac{L'}{L}\right)^2 \quad \text{and} \quad \frac{n'}{n} = \frac{U'}{U}\frac{L}{L'}. \tag{2.10}$$

The relations for the scale factors are

$$P' = M_P\, P, \quad L' = M_L\, L, \tag{2.11}$$

$$U' = M_U\, U, \quad n' = M_n\, n, \tag{2.12}$$

and

$$\rho' = M_\rho\, \rho, \tag{2.13}$$

so that one obtains from Eq. (2.9)

$$\frac{M_P}{M_\rho\, M_U^3\, M_L^2} = 1 \quad \text{and} \quad \frac{M_U}{M_n\, M_L} = 1. \tag{2.14}$$

In this example, there are $n = 5$ physical variables, the rank of the dimension matrix is $r = 3$, and the problem is described by $d = 2$ dimensionless products. So we can define $n - d = r = 3$ scale factors at our discretion, and the remaining scale factors can then be uniquely calculated from Eq. (2.14). If both model and original are operated with the same flow medium (e.g. air), it follows that

$$M_\rho = 1. \tag{2.15}$$

Two more scale factors, e.g. M_L and M_U can still be freely chosen, and the two remaining factors are then obtained to be

$$M_P = M_U^3\, M_L^2 \quad \text{and} \quad M_n = \frac{M_U}{M_L}. \tag{2.16}$$

Complete similarity, that is, the equality of all dimensionless products in model and original, cannot be realized in most cases. Almost always, the geometric similarity is violated, for example, because the relative roughness $k_R\,/\,L$ (with k_R as the average roughness and L as the characteristic length) cannot be kept constant in model and original. The same is true in turbomachinery for the ratio of blade gap to blade height. The problems caused by incomplete geometric similarity and the resulting differences between the behavior of the model and the original are denoted as scale effects.

2.3 Incomplete Similarity

If, apart from these scale effects, complete similarity cannot be achieved anyway, and one or more dimensionless products in model and original cannot be kept at the same value, one speaks of "incomplete similarity". As an example, the flow under simultaneous influence of gravity and friction forces or the determination of stresses and deformations in elastic components under the simultaneous influence of single and weight forces are mentioned here. When complete similarity cannot be achieved and only partially similar experiments can be carried out, additional considerations have to be incorporated that cannot be discussed in general terms, but only for specific tasks.

As an example we consider a ship whose flow resistance W has to be determined. The ship is characterized by its length L and moves with the speed U. It is subject to resistance as it moves through a viscous fluid, e.g. water. The viscosity of the liquid shall be η its density is ρ. A part of the necessary energy for locomotion of the ship is transferred to the wave motion of the liquid. The wave motion is affected by gravity, so that the acceleration due to gravity g must also be included as a variable. One thus obtains

$$W = f\,(U,\, L,\, \eta,\, \rho,\, g). \tag{2.17}$$

Now $n = 6$ and $r = 3$ so that the problem is described by $d = 3$ dimensionless products, and one obtains

$$\frac{W}{\rho\, U^2\, L^2} = f\left(\frac{UL}{\nu},\, \frac{U^2}{gL}\right), \tag{2.18}$$

where $\nu = \eta\, /\, \rho$ denotes the kinematic viscosity. For brevity, we introduce the drag coefficient c_w, the Reynolds number Re and Froud number Fr according to

$$c_w = \frac{W}{\rho\, U^2\, L^2}, \quad \mathrm{Re} = \frac{UL}{\nu} \quad \text{and} \quad \mathrm{Fr} = \frac{U^2}{gL}. \tag{2.19}$$

In order to achieve complete similarity, all dimensionless products for the model and for full-scale must be equal, thus

$$\mathrm{Re} = \mathrm{Re}', \quad \mathrm{Fr} = \mathrm{Fr}' \quad \text{and} \quad c_w = c_w'. \tag{2.20}$$

One obtains

$$\frac{UL}{\nu}\frac{M_U\, M_L}{M_\nu} = \frac{U'L'}{\nu'} \quad \text{respectively} \quad \frac{M_U\, M_L}{M_\nu} = 1, \tag{2.21}$$

and

$$\frac{U^2}{g\,L}\frac{M_U^2}{M_g\,M_L} = \frac{U'^2}{g\,L'} \quad \text{respectively} \quad \frac{M_U^2}{M_L} = 1, \tag{2.22}$$

where because of $g = g'$ the scale factor $M_g = 1$ is already set. If full scale ship and model are both operated in water also $M_\nu = 1$ is fixed and one gains

$$M_U\,M_L = 1 \quad \text{and} \quad \frac{M_U^2}{M_L} = 1 \tag{2.23}$$

and thus immediately

$$M_L = 1 \quad \text{and} \quad M_U = 1 \tag{2.24}$$

that is, model and full-scale ship must be of the same size. Thus, when wanting to perform the experiment on a scaled model, complete similarity has to be abandoned. In the present example the similarity in the Reynolds number is disregarded, and it is thus assumed that

$$c_w = c_{wR}(\text{Re}) + c_{wW}(\text{Fr}). \tag{2.25}$$

In this approach it is assumed that both resistance components are decoupled from each other, which indeed represents a good approximation, but is of course not exactly correct. If now

$$\text{Fr} = \text{Fr}', \tag{2.26}$$

and the scale factors M_U and M_L can be freely selected, it follows that

$$c_{wW}(\text{Fr}) = c'_{wW}(\text{Fr}') \quad \text{respectively} \quad c_{wW} = c'_{wW}. \tag{2.27}$$

The resistance coefficient of the model becomes

$$c'_w = c'_{wW}(\text{Fr}') + c'_{wR}(\text{Re}') \tag{2.28}$$

and the full size version applies to

$$c_w = c_{wW}(\text{Fr}) + c_{wR}(\text{Re}). \tag{2.29}$$

From the difference of Eqs. (2.28) and (2.29) one obtains

$$c_w - c'_w = \left(c_{wW}(\text{Fr}) - c'_{wW}(\text{Fr}')\right) + \left(c_{wR}(\text{Re}) - c'_{wR}(\text{Re}')\right) \tag{2.30}$$

and from this, because of (2.27)

$$c_w - c'_w = \left(c_{wR}(\text{Re}) - c'_{wR}(\text{Re}')\right) \tag{2.31}$$

respectively

$$c_w = c'_w + \left(c_{wR}(\text{Re}) - c'_{wR}(\text{Re}')\right). \tag{2.32}$$

To achieve further progress, one must find an approach to include an analytical calculation of $c_{wR} = c_{wR}(\text{Re})$ or $c'_{wR} = c'_{wR}(\text{Re}')$. If the Reynolds numbers of model and full scale are so large that the flow can be assumed to be fully turbulent in both cases, the frictional resistance of a body can be determined by good approximation by

$$c_{wR} = \text{const} \cdot \text{Re}^{-m} \quad \text{with} \quad m \cong 0,2. \tag{2.33}$$

Thus $c_{wR}(\text{Re})$ as well as $c'_{wR}(\text{Re}')$ are available as analytical functions, and the resistance of the full size version can be determined from

$$c_w = c'_w - c'_{wR}(\text{Re}')\left(1 - \frac{c_{wR}(\text{Re})}{c'_{wR}(\text{Re}')}\right) \tag{2.34}$$

respectively

$$c_w = c'_w - c'_{wR}(\text{Re}')\left(1 - \left(\frac{\text{Re}'}{\text{Re}}\right)^m\right). \tag{2.35}$$

The procedure described here for the determination of the resistance at incomplete similarity is called “up-scaling” and Eq. (2.35) represents a simple form of this procedure.

Chapter 3
Illustrative Examples

This chapter provides several illustrative examples on how to use dimensional analysis in order to obtain the structure of a solution for problems of various areas. The method is applied in this chapter for examples from mechanics, fluid mechanics, thermodynamics, electrical machines, biology,... This shows nicely how powerful and general dimensional analysis can be applied.

3.1 Examples from Mechanics

In the following, we want to apply dimensional analytical methods to problems in the context of mechanics. We limit ourselves to linear elastic deformations, at which the stresses occurring increase proportionally with strain. The proportionality between stress and strain is given by the modulus of elasticity E and Poisson's ratio ν. The quantity ν describes the negative ratio of transverse to axial strain and is already a dimensionless product that shall be respectively included in the following examples, but for the sake of clarity is not explicitly repeated every time.

3.1.1 Large Elastic Deformations

In the theory of elasticity deformations are denote as "large" if the relationship between stress and deformation is no longer linear. This is for example the case for the geometric nonlinearity arising due to large rotations, as for example occurring in fishing rods. We first seek the deformation w of an elastic part under the influence of a single force P and under the forcing of the component's own weight. The general relationship between the deformation w and the characteristic length L of the component, the force P, the modulus of elasticity E, the density ρ and the gravitational acceleration g is then of the form

V. Simon et al., *Dimensional Analysis for Engineers*,
Mathematical Engineering, DOI 10.1007/978-3-319-52028-5_3

$$w = f(L, P, E, \rho, g), \tag{3.1}$$

and the corresponding dimension matrix is

	w	L	P	E	ρ	g
L	1	1	1	−1	−3	1
M	0	0	1	1	1	0
T	0	0	−2	−2	0	−2

From this the following three dimensionless products can be obtained

$$\prod_1 = \frac{w}{L}, \quad \prod_2 = \frac{P}{E L^2} \text{ and } \prod_3 = \frac{\rho g L}{E}. \tag{3.2}$$

Therefore Eq. (3.1) turns into

$$\frac{w}{L} = f\left(\frac{P}{E L^2}, \frac{\rho g L}{E}\right). \tag{3.3}$$

Asking for the strain, then a corresponding relation to Eq. (3.1) of the type

$$\sigma = f(L, P, E, \rho, g) \tag{3.4}$$

must exist. The dimension matrix is now given by

	σ	L	P	E	ρ	g
L	−1	1	1	−1	−3	1
M	1	0	1	1	1	0
T	−2	0	−2	−2	0	−2

and in a first step one obtains the dimensional products

$$\prod_1 = \frac{\sigma L^2}{P}, \quad \prod_2 = \frac{P}{E L^2} \text{ and } \prod_3 = \frac{\rho g L}{E}. \tag{3.5}$$

Thus, Eq. (3.4) can be rewritten as

$$\frac{\sigma L^2}{P} = f\left(\frac{P}{E L^2}, \frac{\rho g L}{E}\right). \tag{3.6}$$

The Eqs. (3.3) and (3.6) assume an elastic material behavior, but apart from that apply in general. If the structure is not loaded by a single force, but by a moment M, the load P can be replaced by M/L.

3.1.2 Small Elastic Deformations

For the case of small elastic deformation, Eqs. (3.3) and (3.6) can be further simplified. “Small” deformations imply that the problem under consideration can be described sufficiently accurate by the linear theory of elasticity. In that framework, the deformation w is directly proportional to the external load P. First we neglect the deformation due to the part’s own weight and then from Eq. (3.3) we obtain the sharpened statement

$$\frac{w}{L} = \frac{P}{E\,L^2}\ \text{const.} \tag{3.7}$$

respectively

$$\frac{w\,E\,L}{P} = \text{const.} = C_1. \tag{3.8}$$

If the deformation is solely caused by the own weight of the structure, because of $P \propto \rho\,g\,L^3$ in Eq. (3.7) the force P is replaced by $\rho\,g\,L^3$ and the relation takes the form

$$\frac{w}{L} = \frac{\rho\,g\,L}{E}\ \text{const.} \tag{3.9}$$

respectively

$$\frac{w\,E}{\rho\,g\,L^2} = \text{const.} = C_2. \tag{3.10}$$

Since in the framework of linear elasticity theory, the deformation is directly proportional to the external load, the deformation at simultaneous loading by individual forcing and own weight can be obtained by superposition, thus

$$\frac{w\,E\,L}{P} = C_1 + C_2\,\frac{\rho\,g\,L^3}{P}. \tag{3.11}$$

Accordingly from Eq. (3.6) the case of stress due to a single force results to

$$\frac{\sigma\,L^2}{P} = \text{const.} = C_1, \tag{3.12}$$

and for the stress under load by the part’s own weight due to $P \propto \rho\,g\,L^3$

$$\frac{\sigma}{\rho\,g\,L} = \text{const.} = C_2. \tag{3.13}$$

Again, because of the assumed linearity for the case of simultaneous loading by single forcing as well as the part's own weight force the relationship is determined via superposition to be

$$\frac{\sigma L^2}{P} = C_1 + C_2 \frac{\rho g L^3}{P}. \tag{3.14}$$

3.1.3 Deformation of an Elastic Beam

In many practical applications elastic bodies can be idealized by a beam. A beam is understood as a body, whose typical axial length L is much larger than the length measured perpendicularly to the expansion axis (see Fig. 3.1). In these cases, the product of the modulus of elasticity and the second area moment J has to be considered together, so that Eq. (3.1) for the elastic deformation of a beam under the effect of a single force takes the form

$$w = f(L, P, EJ). \tag{3.15}$$

From the dimension matrix

	w	L	P	EJ
L	1	1	1	3
M	0	0	1	1
T	0	0	−2	−2

one obtains the dimensionless products

$$\prod\nolimits_1 = \frac{w}{L} \quad \text{and} \quad \prod\nolimits_2 = \frac{P L^2}{E J}. \tag{3.16}$$

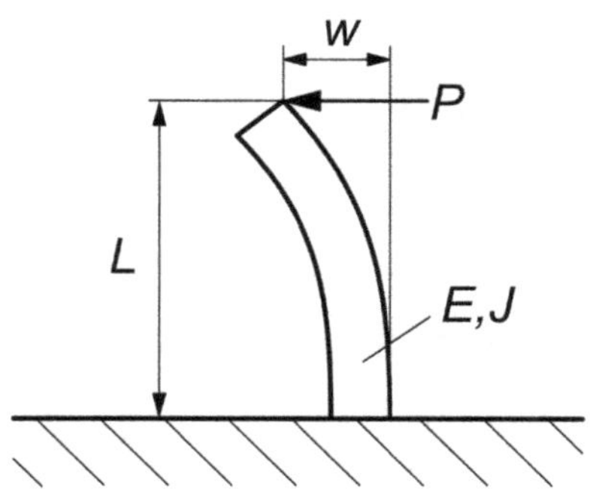

Fig. 3.1 Deformation of an elastic beam

This leads to the relation

$$\frac{w}{L} = f\left(\frac{PL^2}{EJ}\right). \tag{3.17}$$

Due to the proportionality of w and P for linear elastic deformations it follows

$$\frac{w}{L} = \text{const.}\ \frac{P\,L^2}{E\,J} \tag{3.18}$$

hence

$$\frac{w\,E\,J}{P\,L^3} = \text{const.} \tag{3.19}$$

For a one sided fixed and loaded beam, with a cylindrical cross section that is loaded at the free end (Fig. 3.1), the theory provides a value for the constant of $1/3$ (see e.g. Szabo 1975).

We now consider the special case of a beam with a circular cylindrical cross section with the diameter d, which is loaded by its own weight. Then because of $P \propto \rho\, g\, d^2\, L$ and $J \propto d^4$ we obtain from Eq. (3.17) the special form

$$\frac{w}{L} = f\left(\frac{\rho\, g\, L^3}{E\, d^2}\right), \tag{3.20}$$

which also is still valid for large elastic deformations. From Eq. (3.20) some very interesting conclusions can be drawn. In trees, for example, the ratio ρ/E is approximately constant and independent of the size of the trees. If we now assume that trees of different types and sizes grow so that the relative deformation w/L due to the load is always the same to provide for adequate protection against breakage, yields

$$\frac{\rho\, g\, L^3}{E\, d^2} = \text{const.} \tag{3.21}$$

Thus, the stem diameter d and the tree height L are related by

$$d \propto L^{3/2}. \tag{3.22}$$

Because of $d^2 L \propto m$ one obtains

$$L \propto m^{1/4} \quad \text{and} \quad d \propto m^{3/8}. \tag{3.23}$$

In comparison, in terms of geometric similarity with $d \propto L$ and $L^3 \propto m$ one would obtain the relationships

$$L \propto m^{1/3} \quad \text{and} \quad d \propto m^{1/3}. \tag{3.24}$$

Thus, with increasing mass the diameter increases faster than the height under the conditions of elastic similarity. Trees are therefore not geometrically, but elastically similar. The same reasoning can also be applied to living beings. For growing people, for example, the typical leg, arm and torso diameter increases faster than the length of the corresponding body parts. In principle it can thus be explained why large and small mammals are not geometrically, but elastically similar: In the case that a mouse would be geometrically similar enlarged to the size of an elephant, its legs would break as a result of its own weight. In elastic similarity the leg diameter grows faster than it would for geometric similarity, so that large creatures have the necessary statics. The fact that large organisms also have more strength than expected by geometric scaling can be explained by the fact that due to the cell structure the muscle fibers are independent of the size, and muscle strength thus grows directly proportional to the diameter of the muscle (Thompson 1917). The muscle strength of elastically similar animals is therefore

$$F_{muscle} \propto d^2 \propto m^{3/4}, \tag{3.25}$$

whereas in the case of geometric similarity it would follow to

$$F_{muscle} \propto d^2 \propto m^{2/3} \tag{3.26}$$

Also other physiological measures of living beings are proportional to $m^{3/4}$. This law was first formulated by Max Kleiber and is thus known as the "Law of Kleiber" (see e.g. Mc Mahon and Bonner 1985).

3.1.4 Critical Load and Buckling of a Beam

As another example, it is asked for the critical load acting on a beam in axial direction, just such that the beam remains stable before kinking to the side (Fig. 3.2). This critical load is also called buckling load and for thin beams is a function of the beam length and the product of the elasticity modulus and the moment of inertia, so that

$$P = f(EJ, L). \tag{3.27}$$

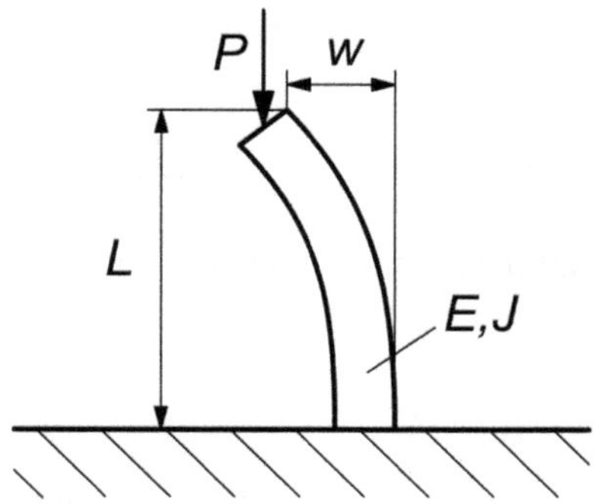

Fig. 3.2 Buckling load of a beam

We obtain the single dimensionless product

$$\frac{P\,L^2}{E\,J} = \text{const.} \tag{3.28}$$

For a one-sided fixed beam, the theory provides the value $\pi^2/4$ for the constant in Eq. (3.28), Greenhill (1881). Large upright beams, such as flagpoles or chimneys may buckle under their own weight. For circular cylindrical beams under the loading of the own weight from Eq. (3.28) by application of $P \propto \rho\, g\, d^2\, L$ and $J \propto d^4$ again Eq. (3.21) is obtained. So, to explain the geometric ratios of living beings, instead of the demand for the same relative deformation due to some load, by Eq. (3.21) one might also yield formal reasoning by the requirement for sufficient stability to a critical buckling load.

3.1.5 Natural Frequency of an Elastic Rod

As a final example from mechanics the natural frequency of an elastic rod is regarded. The rod has the elasticity modulus E, the density ρ, the length L and diameter d. The natural frequency n of the rod is thus a function of the form

$$n = f(E,\, \rho,\, L,\, d). \tag{3.29}$$

Therefore the dimension matrix is obtained as

	n	E	ρ	L	d
L	0	−1	−3	1	1
M	0	1	1	0	0
T	−1	−2	0	0	0

The dimension matrix yields the two dimensionless products

$$\prod_1 = \frac{L}{D} \quad \text{and} \quad \prod_2 = n\,L\sqrt{\frac{\rho}{E}} \tag{3.30}$$

and the relation equivalent to Eq. (3.29) follows to

$$n\,L\sqrt{\frac{\rho}{E}} = f\left(\frac{L}{D}\right). \tag{3.31}$$

An even clearer result is obtained when considering that in the context of the beam theory, the special cross sectional shape, here the diameter d, only enters through the area moment of inertia J, and that only the product $E\,J$ is important. When additionally applying the mass m instead of the density ρ, it follows that the natural frequency must obey a functional relation of the form

$$n = f(EJ,\, m,\, L) \tag{3.32}$$

The dimension matrix in this case is given by

	n	EJ	m	L
L	0	3	0	1
M	0	1	1	0
T	−1	−2	0	0

and the only dimensionless product is

$$n\,L\sqrt{\frac{m\,L}{EJ}} = \text{const.} \tag{3.33}$$

Accounting for $J \propto d^4$ and $m \propto \rho\, d^2\, L$ yields

$$n\,\frac{L^2}{d}\sqrt{\frac{\rho}{E}} = \text{const.} \tag{3.34}$$

Thus, the natural frequency is inversely proportional to the square of the length and directly proportional to the diameter. These predictions can be easily illustrated by pushing a rod onto the edge of a table, so that the end protrudes. Plucking on the protruding part produces a buzzing sound due to the oscillation. Someone with a musical ear can easily hear that at halving the protruding end the sound becomes two octaves higher, as the frequency quadruples.

3.2 Examples from Fluid Mechanics

3.2.1 Surface Waves

As a first example, we consider surface waves along the interface of an incompressible liquid and ask for the phase velocity c of these waves, e.g. the speed with which the wave crests move forward. We neglect the influence of viscosity and restrict the consideration to the case of small amplitudes (i.e., the amplitudes are small compared to all other geometrical dimensions). Besides, the properties of the gaseous phase at the interface are not considered. Depending on the restoring force it is generally distinguished between gravity waves and capillary waves, Spurk and Aksel (2010).

At first **gravity waves** in shallow water of the depth h are regarded. The relationship between the phase velocity and the other physical quantities g, h and ρ is then

$$c = f(g, \rho, h). \tag{3.35}$$

Arranging the variables in a dimension matrix yields

	C	g	ρ	h
L	1	1	−3	1
M	0	0	1	0
T	−1	−2	0	0

Since the density is the only quantity that contains the mass as a dimension, one must conclude that it cannot enter the functionality. The remaining variables enter the single dimensionless product

$$\frac{c}{\sqrt{gh}} = \text{const.} \quad \text{respectively} \quad c = \pm\text{const.}\sqrt{gh}. \tag{3.36}$$

The phase velocity of the waves therefore depends on the water depth and decreases with increasing depth. Strictly speaking Eq. (3.36) is valid only for waves at constant depth. Yet, still applying Eq. (3.36) locally to the case of decreasing water depth, for instance close to the shore, and denoting the height of a wave crest by h_1 and the height of the wave base by h_2 yields the phase velocity ratio

$$\frac{c_1}{c_2} = \sqrt{\frac{h_1}{h_2}}, \tag{3.37}$$

and it becomes apparent that wave crests propagate faster than the wave base, allowing to qualitatively comprehend the wave breaking at the beach. Applying

Eq. (3.36) to waves of large depth, such as ocean waves, yields the unrealistic conclusion that these waves travel at very high speed. In this case, the water depth therefore does not represent a suitable length for the dimensionless product. However, the only other characteristic length is the wavelength λ, so that for $h/\lambda \gg 1$ the phase velocity results to

$$c = \pm\text{const.}\sqrt{g\lambda}. \tag{3.38}$$

Waves with a phase velocity in accordance to Eq. (3.38) are typically referred to as "**long gravity waves**".

So far we have only considered gravity waves, i.e. wave motion that is significantly influenced by gravity. The interface between the two fluids is always affected by the surface tension σ. Under silent conditions, the pressure difference is proportional to the surface tension and the mean curvature of the surface. A wave on the surface causes a pressure difference across the surface, inducing a motion of the adjacent fluid. We therefore suspect, that the surface tension can cause wave motion, and the regarded phase velocity takes a functional relationship of the form

$$c = f(g, \rho, \lambda, \sigma). \tag{3.39}$$

The dimension matrix then takes the form

	c	g	ρ	λ	σ
L	1	1	−3	1	0
M	0	0	1	0	1
T	−1	−2	0	0	−2

from which the two dimensionless products

$$\Pi_1 = \frac{c}{\sqrt{g\lambda}} \quad \text{and} \quad \Pi_2 = \frac{\sigma}{\rho g \lambda^2} \tag{3.40}$$

are obtained. The limiting case that the wave motion is not driven by gravity g at all, but rather by surface tension only can be studied more easily when incorporating only one single dimensionless product. We therefore rearrange

$$\Pi_1' = \Pi_1 \Pi_2^{-1/2} = c\sqrt{\frac{\rho\lambda}{\sigma}} \tag{3.41}$$

and obtain

$$c\sqrt{\frac{\rho\lambda}{\sigma}} = f\left(\frac{\sigma}{\rho g \lambda^2}\right). \tag{3.42}$$

The quantity

$$a = \sqrt{\sigma/(\rho g)} \tag{3.43}$$

has the dimension of a length and is referred to as "**Laplacian length**", Spurk and Aksel (2010). For $a/\lambda \to \infty$ gravity has a vanishing influence, and Eq. (3.42) leads to

$$c = \text{const.}\sqrt{\frac{\sigma}{\rho\lambda}}. \tag{3.44}$$

The described waves are called capillary waves. As in the case for long gravity waves they are so called dispersive waves, i.e. their phase velocity depends on the wavelength. However, the phase velocity of capillary waves increases as the wavelength decreases (abnormal dispersion), while the phase velocity of gravity waves increases, when the wavelength is increased (normal dispersion). Therefore, there exists a minimum of the phase velocity. From an analytical consideration of the problem, Spurk and Aksel (2010) one obtains

$$c = \sqrt{\frac{2\pi\sigma}{\rho\lambda} + \frac{g\lambda}{2\pi}}. \tag{3.45}$$

For the properties of water and air the smallest possible phase velocity then results to be $c \approx 0.23\text{m/s}$.

3.2.2 *Ship Waves*

A particularly beautiful and rich wave system can be observed when regarding ships or also ducks swimming on a lake. Following Epshteyn (1977), we first consider water surface waves caused by a local disturbance and seek for the location of constant phase, e.g. the distance r of a wave crest to the origin of the disturbance after a certain time t. As in the last example, we neglect the viscosity of the liquid, since it obviously cannot explain the regarded phenomenon. Furthermore, the water depth is assumed to be large in comparison with the wavelength. The desired relation is therefore of the form

$$r = f(t, \rho, g). \tag{3.46}$$

As before, the density cannot enter the problem, and the only dimensionless parameter is

$$\frac{r}{g\,t^2} = \varphi = const. \tag{3.47}$$

Equation (3.47) reveals that the distance of the considered wave crest changes with the square of time, and therefore, the phase velocity increases with time. When throwing a stone into a lake, it is observed that a large number of wave crests move outwards with increasing speed from the impact location.

If one would regard a different phase, thus a different wave crest, at fixed time t, one would again obtain Eq. (3.47). It is therefore apparent that the constant φ indicates the phase of the caused disturbance, thus takes different values for different phases.

We now regard the disturbance caused by a moving ship that moves in a straight line with a constant speed U, as described by Eq. (3.47). In time t, in which the ship travels the distance $x_0(t) = Ut$, the perturbation spreads evenly in all directions. After the duration t, from the view of an observer placed on the ship, the disturbances initiated at $t = 0$ form a circle (see Fig. 3.3), whose distribution is determined by the equation

$$F \equiv (x - x_0(t))^2 + y^2 - r^2(t) = 0. \tag{3.48}$$

Along the boat path at each location $x_0(t)$, that is, at any time t such a disturbance is caused. Since the propagation of the disturbances is finite, a curve that envelopes all disturbances can be specified. This envelope is calculated by $F = 0$ and $\partial F/\partial t = 0$ in implicit form, Spurk (1992) as

$$x = Ut\left(1 - 2\left(\frac{\varphi g t}{U}\right)^2\right) \tag{3.49}$$

and

$$y = \pm Ut\frac{\varphi g t}{U}\sqrt{1 - 4\left(\frac{\varphi g t}{U}\right)^2}. \tag{3.50}$$

For different phases φ by Eqs. (3.49) and (3.50) different curves are described; φ thus is a parameter for the set of curves (see Figs. 3.4 and 3.5), and t is a curve

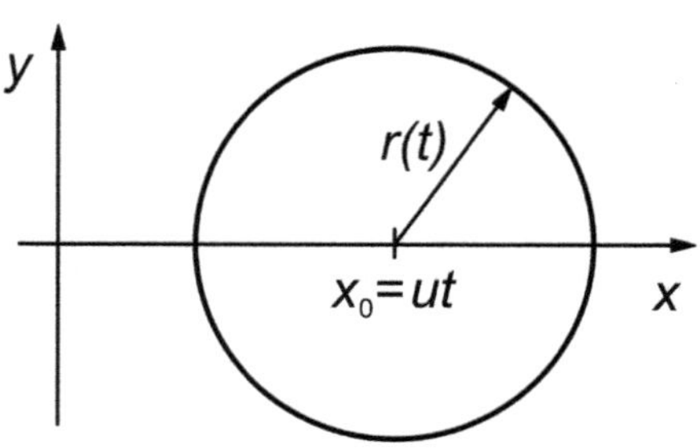

Fig. 3.3 Perturbation at time t

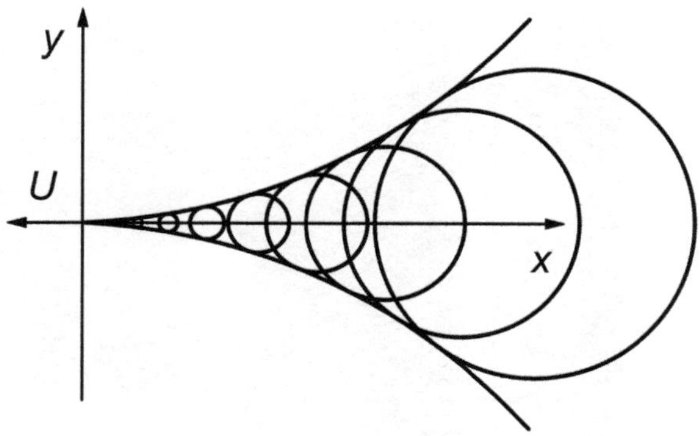

Fig. 3.4 Envelope of disturbance caused at different times of the same phase. The figure is based on Spurk (1992)

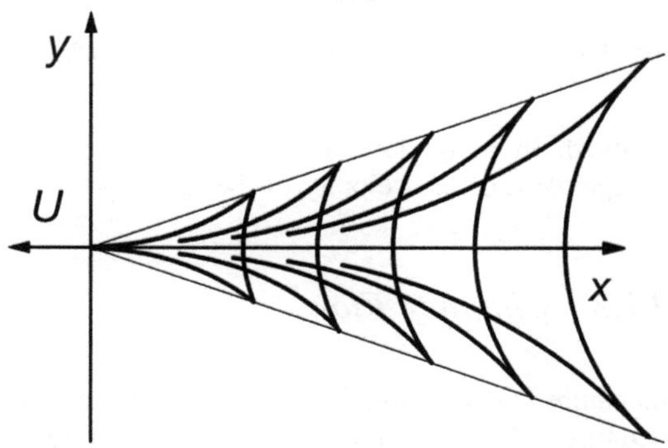

Fig. 3.5 Wave system of a ship moving at constant speed (envelopes for disturbances of different phases). The figure is based on Spurk (1992)

parameter. For $t = U/(\varphi g\sqrt{6})$ each of these envelopes reaches maximum coordinate values x_m, y_m determined by

$$x_m = \frac{2}{3\sqrt{6}}\frac{U^2}{\varphi g} \tag{3.51}$$

and

$$y_m = \frac{1}{6\sqrt{3}}\frac{U^2}{\varphi g}. \tag{3.52}$$

The ratio $y_m/x_m = 1/\sqrt{8}$ is surprisingly independent of the phase of the perturbations. Each of these envelopes touches two straight lines intersecting at the origin of the coordinate system, where the ship is located, and moving with the ship. The opening angle, enclosed by these lines is 2 × 19.47° = 38.94°. Thus, along the water line a finite disturbance of the surface is even expected, if the momentum of the underlying impact is neglected as in the present example. In fact, in the case of a ship moving at constant speed, as well as in the case of ducks swimming on a lake, the wave system is observed to be bounded by two straight lines that are particularly

Fig. 3.6 Wave system shown for a swimming duck (Wikipedia)

evolved. In particular, the opening angle of the wave system bounding lines is independent of the velocity of the moving body (Fig. 3.5 and Fig. 3.6).

3.2.3 Chimney Flow

In nature and technology often flows are encountered, which are driven by temperature differences and thus by buoyancy forces. As a simple example we consider the flow through a chimney of height h, which is sketched in Fig. 3.7. The flow is driven by the temperature difference

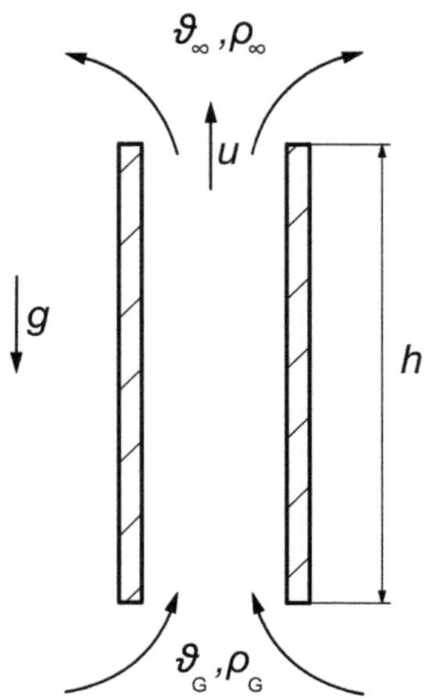

Fig. 3.7 Buoyancy flow through a chimney

$\Delta\vartheta = \vartheta_G - \vartheta_\infty$, where ϑ_G denotes the temperature of the hot gas at the bottom of the chimney and ϑ_∞ represents the temperature of the undisturbed ambient. With increasing temperature, the density decreases, and specifically lighter particles are moving upwards, opposite to gravity. At this the density changes with temperature are crucial for the problem. If the temperature variations are not too large, the pressure remains approximately constant, and the change of density with temperature at constant pressure is determined by the coefficient of thermal expansion, Kays et al. (2004)

$$\beta = -\frac{1}{\rho}\left(\frac{\partial \rho}{\partial \vartheta}\right)_p, \tag{3.53}$$

that takes the value $\beta = 1/\vartheta_\infty$ for ideal gas. Because gas particles, with a temperature difference to the ambient experience a force only under the action of gravity, the force of gravity g occurs only in the combination $g\beta$. Now we ask for the speed u with which hot gas emerges from a chimney of height h, and suggest a relationship of the form

$$u = f(h, g\beta, \Delta\vartheta, \rho_\infty). \tag{3.54}$$

That yields the dimension matrix

	u	h	$g\beta$	$\Delta\vartheta$	ρ
L	1	1	1	0	−3
M	0	0	0	0	1
T	−1	0	−2	0	0
θ	0	0	−1	1	0

Again the density can obviously not enter the problem, and the only dimensionless product

$$\frac{u}{\sqrt{hg\beta\Delta\vartheta}} = \text{const.} \tag{3.55}$$

is obtained. The velocity thus increases proportionally to the square root of the height of the chimney and proportional to the square root of the temperature difference. It is noteworthy that in fluids under the influence of gravity and a temperature difference always a speed of the order $\sqrt{hg\beta\Delta\vartheta}$ occurs (see Kays et al. 2004). It thus initiates a fluid motion, even if no other peripheral driving force is present. For example that is the case near heated surfaces, but also in gravity currents along the seabed when cold water masses push warmer water and move along the seabed, often with considerable speed.

3.2.4 *Hydraulic Machines*

Dimensional analytical methods are widely used due to their major benefits for all types of turbomachinery. The benefits of dimensional analytical methods are tremendous, as they allow substantial cost reductions for expensive test machines. As the simplest example we consider hydraulic machinery such as pumps, fans and turbines, where the influence of compressibility of the fluid can be neglected. In order to have a specific application in mind, we think of a pump that is driven by the rotational speed Ω to promote the flow rate $\dot{V}$. The pump generates a pressure increase Δp in the fluid. The fluid is characterized by its viscosity η and density ρ, while the machine is characterized by a characteristic length, such as the rotor diameter d. All other geometric length scales of the machine are non-dimensionalised by the characteristic length, and therefore incorporate additional dimensionless products, that are not listed here as variables, so that we consider geometrically similar machines (Fig. 3.8). If there are no free surfaces, the pressure difference generated by the engine can be expressed in the form

$$\Delta p = f(\dot{V}, \Omega, d, \eta, \rho), \tag{3.56}$$

and the dimension matrix becomes

	Δp	$\dot{V}$	η	Ω	d	ρ
L	−1	3	−1	0	1	−3
M	1	0	1	0	0	1
T	−2	−1	−1	−1	0	0

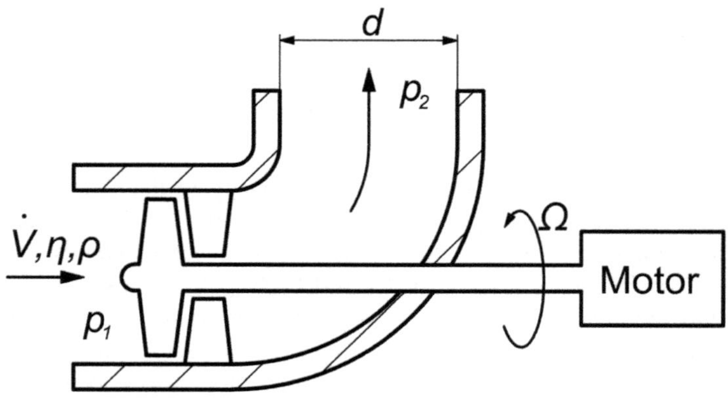

Fig. 3.8 Schematics of an axial pump

The rank of this matrix is three, so that three dimensionless products are expected for the six variables that follow after transformation to

$$\prod_1 = \frac{\Delta p}{\rho \Omega^2 d^2}, \quad \prod_2 = \frac{\dot{V}}{\Omega d^3} \quad \text{and} \prod_3 = \frac{\rho \Omega d^2}{\eta} = \mathrm{Re}, \tag{3.57}$$

where $\prod_3$ has already been identified as a form of the Reynolds number. Equation (3.56) transforms to

$$\frac{\Delta p}{\rho \Omega^2 d^2} = f\left(\frac{\dot{V}}{\Omega d^3}, \mathrm{Re}\right). \tag{3.58}$$

The Reynolds number is not a measure of mechanisms that explain the operation of the machine; it only exerts an influence on the operation of the machine. Often, the Reynolds number is so large that its influence is negligible. By doing so, we neglect the effect of viscosity and thus friction forces in the flow. In these cases, one obtains

$$\frac{\Delta p}{\rho \Omega^2 d^2} = f\left(\frac{\dot{V}}{\Omega d^3}\right). \tag{3.59}$$

The number of variables is therefore reduced to only two variables, and it obviously suffices to apply a single curve, in order to fully describe the operating range of a machine.

3.3 Examples from Thermodynamics

3.3.1 *Thermal Equation of State for Ideal Gases*

The pressure which gas exerts on the walls of a vessel is caused by the collisions of the gas molecules with the vessel wall. We therefore assume that the pressure p depends on the mass m of the molecules, of the average molecular velocity v and on the number of molecules n per unit volume so that the relation is written in the form

$$p = f(m, v, n). \tag{3.60}$$

The dimension matrix then reads

	p	m	v	n
L	−1	0	1	−3
M	1	1	0	0
T	−2	0	−1	0

and the dimensionless product

$$\frac{p}{mnv^2} = \text{const.} = C_1 \tag{3.61}$$

respectively

$$p = C_1 mnv^2 \tag{3.62}$$

is obtained. According to the kinetic theory of gases (see e.g. Vincenti and Kruger 1986) the temperature of a gas is proportional to the average kinetic energy of the molecules, thus

$$\vartheta k_B = \left\langle \frac{mv^2}{2} \right\rangle, \tag{3.63}$$

where k_B denotes the Boltzmann constant. In addition, the density is proportional to mn so that Eq. (3.62) takes the form

$$p = C_2 \frac{k_B\, \rho\, \vartheta}{m}, \tag{3.64}$$

in which C_2 represents a constant. The mass m is in turn proportional to the molecular weight M of the molecules, because the molecular weight is defined as 32 times the ratio of the mass of the regarded gas molecules to the mass of an oxygen molecule, thus

$$M = 32 \frac{m}{m_{O_2}}. \tag{3.65}$$

Thus Eq. (3.64) finally takes the form

$$p = \frac{\rho R \vartheta}{M}, \tag{3.66}$$

where

$$R = C_3\, k_B / m_{O_2}, \qquad C_3 = \text{const.} \tag{3.67}$$

is a constant for all gases with the dimension $[R] = \mathrm{L}^2\mathrm{T}^{-2}\theta^{-1}$. This constant is the universal gas constant R_m = 8.314 J/(mol K).

3.3.2 Strong Explosion

In his famous work from 1950 G.I. Taylor introduced a solution to the problem of strong explosions, Taylor (1950). At a strong, spherically symmetrical explosion at time $t = 0$ at the location $r = 0$ the energy E is instantaneously released. As a result of the explosion a shock wave is formed, whose radius r increases over time t. In general, the problem is influenced by the density ρ_0 and the pressure p_0 in the undisturbed ambient. We now ask for the location of the shock front and thus regard the relationship

$$r = f(t, \rho_0, p_0, E). \tag{3.68}$$

This gives rise to the dimension matrix

	r	t	ρ_0	p_0	E
L	1	0	−3	−1	2
M	0	0	1	1	1
T	0	1	0	−2	−2

After rearranging we obtain the sought relation as

$$f\left(\frac{E\,t^2}{\rho_0\, r^5}, \frac{p_0\, t^2}{\rho_0\, r^2}\right) = 0. \tag{3.69}$$

Based on experience, for high energy explosions a spherical shock wave originating from the explosion center propagates with a very large speed compared to the undisturbed speed of sound. From the shock relations of gas dynamics, Spurk and Aksel (2010), we know that the pressure behind the shock wave is much larger than the undisturbed pressure p_0, so that in the limiting case of strong explosions

$$\frac{p_0\, t^2}{\rho_0\, r^2} \to 0. \tag{3.70}$$

Then from Eq. (3.69) it follows that

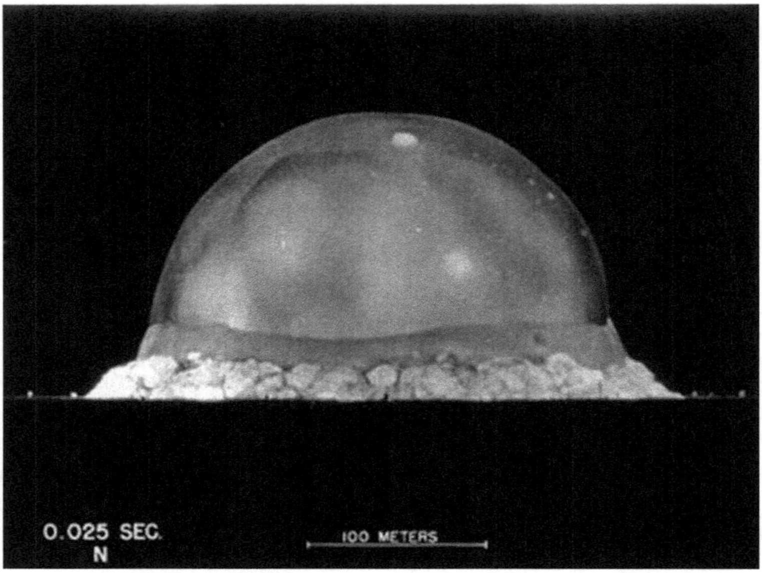

Fig. 3.9 Photograph of the atomic bomb explosion (the Trinity test), that GI Taylor used to estimate the explosion energy (www.trinityremembered.com)

$$\frac{E\,t^2}{\rho_0\,r^5} = \text{const.} \tag{3.71}$$

respectively

$$r = \text{const.}\left(\frac{E\,t^2}{\rho_0}\right)^{1/5}. \tag{3.72}$$

The theory provides the value 1.033 for the constant. The photo (Fig. 3.9) was taken by the U.S. military in 1945 as part of the Trinity tests and published in Life magazine. G.I. Taylor estimated from the photo the radius at time $t = 0.025$ s with $r = 140$ m and assumed $\rho_0 = 1\,\text{kg/m}^3$. Thus, he estimated $E = 90$ TJ. Even though the photo was published freely, the value of the explosive force was strictly confidential.

3.3.3 *Unsteady Heat Transfer*

We now consider a simple case of unsteady heat transfer and think of a body with temperature $\vartheta(t = 0) = \vartheta_0$ at $t = 0$ (Fig. 3.10). The ambient around the body has the temperature ϑ_∞, and we assume that the ambient temperature remains constant in time and space. This assumption is often justified in flow problems around bodies to a good approximation, but also for bodies located in a well-mixed liquid. When the ambient temperature is larger than the temperature of the body, thus $\vartheta_\infty > \vartheta_0$, the body is heated until it will reach ambient temperature asymptotically for large times. For transient temperature problems, the temperature of the body is affected by the heat capacity of the body. Since c is the specific heat capacity, i.e. the amount of heat per unit mass, which is necessary to obtain a temperature change of one degree for 1 kg mass, it is useful to apply the heat capacity $\rho\, c$, that is the thermal capacity per unit volume in the present context. The temperature in the body is also influenced by the thermal conductivity k of the body, the heat transfer coefficient h (which is a measure of the amount of heat transferred to the body), and of course, by the extent of the body characterized by its typical length L. We further assume that heat conduction in the body is so effective that the temperature in the body is homogeneous (i.e. spatially constant). Then it follows that

$$\frac{\vartheta - \vartheta_0}{\vartheta_\infty - \vartheta_0} = f(k,\, h,\, \rho c_p,\, L,\, t), \tag{3.73}$$

and the dimension matrix has the entries

	k	h	ρc	L	t
L	1	0	−1	1	0
M	1	1	1	0	0
T	−3	−3	−2	0	1
Θ	−1	−1	−1	0	0

Thus, after rearranging, the following functional dependence for the time dependent temperature change is obtained to be

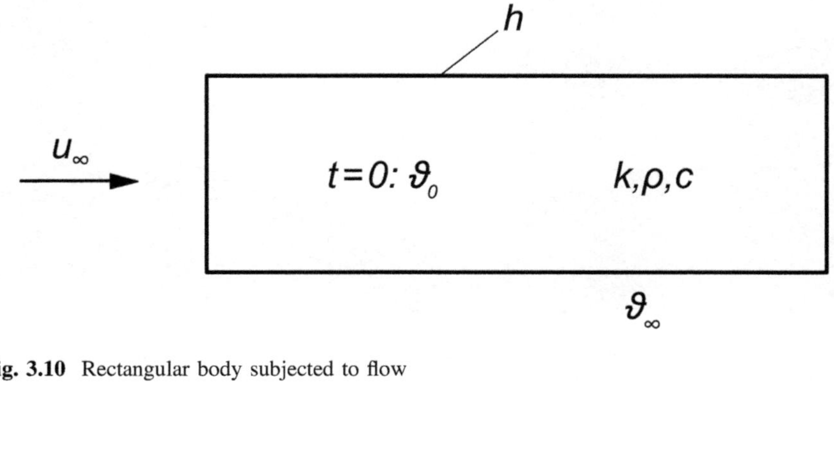

Fig. 3.10 Rectangular body subjected to flow

$$\frac{\vartheta - \vartheta_\infty}{\vartheta_0 - \vartheta_\infty} = f\left(\frac{k\,t}{\rho\,c\,L^2}, \frac{h\,L}{k}\right). \tag{3.74}$$

3.4 Examples from Nature

3.4.1 The Dynamics of Planetary Rings

The ring system of Saturn is one of the most fascinating natural phenomena. The ring system consists of particles that travel in circular orbits around the planet with angular and circumferential velocities (Fig. 3.11)

$$\Omega = \sqrt{\frac{GM}{r^3}} \tag{3.75}$$

and

$$u = \Omega r, \tag{3.76}$$

respectively. Here, G is the gravitational constant, M is the mass of the central planet and r is the radial position of the particle. Particles that are closer to the planet travel at a greater speed, so that two particles located on neighbouring radial positions will eventually collide. This is the source of a fluctuation velocity of the particles, which is superimposed on the mean circumferential flow velocity. Because the particles are inelastic, a fraction of the particle kinetic energy is lost by

Fig. 3.11 Saturn and its rings in natural colors. The picture was taken by the space orbiter Cassini on October 6, 2004 (Wikipedia)

dissipation: in a collision between two particles, their normal velocity components are related through the coefficient of restitution ε. When ε equals one, collisions are perfectly elastic and energy is conserved. Values of ε less than one involve a loss of kinetic energy. It is important to mention that if the particles were ideally elastic, the particle flow would not be confined to a thin ring system. Rather, the particles would consume all available space around the planet. Only a small amount of inelasticity is necessary to confine the particle flow to a thin ring. The dynamics of the ring system is thus governed by the steady balance between collisional productions of the velocity fluctuations as driven by the mean Keplerian shear at a rate

$$\frac{\mathrm{d}u}{\mathrm{d}r} = -\frac{1}{2}\Omega \tag{3.77}$$

and its dissipation in collisions. As the analysis of planetary ring dynamics has successfully employed methods from the classical kinetic theory of gases, the mean fluctuation velocity is also called the "granular temperature" and is denoted by θ, having the dimension of velocity squared. We restrict the analysis to a dilute flow of identical smooth spherical particles with diameter D and density ρ_s. The flow density is denoted by ρ. We now want to determine the magnitude of the granular temperature θ as a function of D, Ω, ε, ρ_s and ρ, and find

$$\theta = f(D, \Omega, \rho, \rho_s, \varepsilon). \tag{3.78}$$

The dimension matrix then reads for this problem

	D	Ω	ρ	ρ_s	ε	θ
L	1	0	−3	−3	0	2
M	0	0	1	1	0	0
T	0	−1	0	0	0	−2

From this, we obtain from dimensional analysis that three dimensional products will describe the problem under consideration, thus

$$\frac{\sqrt{\theta}}{D\Omega} = f(\mu, \varepsilon), \tag{3.79}$$

where the solid volume fraction μ has been defined as

$$\mu = \frac{\rho}{\rho_\mathrm{S}}. \tag{3.80}$$

Since θ increases when μ decreases, we can try to expand $(D\Omega)/\sqrt{\theta}$ in terms of μ to get

$$\frac{D\Omega}{\sqrt{\theta}} = f_1(0,\varepsilon) + \mu f_1'(0,\varepsilon) + \frac{\mu^2}{2} f_1''(0,\varepsilon) + \cdots. \tag{3.81}$$

With $f_1(0, \varepsilon) = 0$ and neglecting powers of μ higher than the first, we obtain

$$\frac{\mu\sqrt{\theta}}{D\Omega} = f(\varepsilon). \tag{3.82}$$

With the help of dimensional analysis, we cannot determine the precise form of the function $f(\varepsilon)$. This can only be done by solving the governing equations of the flow. The reader is referred to the paper by Simon and Jenkins (1994) for more details on this. The dimensionless quantity $S = (D\Omega/\mu\sqrt{\theta})$ appears in every granular shear flow and can be interpreted as a non-dimensional shear rate. Of course, the granular temperature as well as the flow density vary across the ring and we give their midplane values (i.e. at z = 0) the index 0. We may now define a ring thickness t_1 by

$$t_1 = \frac{2}{\rho_s} \int_0^\infty \rho \mathrm{d}z, \tag{3.83}$$

where z denotes the axial coordinate normal to the ring plane. Obviously, t_1 is the thickness of the ring if the particles were pressed together to form a densely packed solid disc. Using $\eta = z\Omega/\sqrt{\theta_0}$ as a non-dimensional axial distance, we get

$$\frac{t_1}{D} = 2\frac{\mu_0\sqrt{\theta_0}}{D\Omega} \int_0^\infty \frac{\mu}{\mu_0} d\eta. \tag{3.84}$$

In astronomy, it is more customary to use a quantity called "optical depth" ϕ which is related to t_1 by

$$\phi = \frac{3\,t_1}{2\,D} = \frac{3}{S_0} \int_0^\infty \frac{\mu}{\mu_0} d\eta. \tag{3.85}$$

Another, second measure of the ring's thickness may be defined by integrating the density over z for the whole domain

$$t_2 = \frac{2}{\rho_0} \int_0^\infty \rho\, dz, \tag{3.86}$$

hence

$$\frac{t_2}{D} = \frac{2}{\mu_0 S_0} \int_0^\infty \frac{\mu}{\mu_0} d\eta. \tag{3.87}$$

The thickness t_2 is a rough measure of the apparent visual thickness of the ring, as it is an exact measure of the ring's axial extent if ρ were constant. Of course, the value of S cannot be found from dimensional analysis alone. A detailed analytical treatment of the problem (Simon and Jenkins 1994) provides values of S for a range of values of ε. From there, we find that for ε ranging between 0 and 0.63, μ_0 ranges between 0.003 and 0.009, D ranges from a few centimetres up to 10 m, $\sqrt{\theta_0}$ ranges from 0.002 to 0.005 m/s, ϕ ranges from 1 to 4, and t_2 ranges from 20 m up to 2000 m.

3.4.2 The Flight Velocity of Birds

Dimensional analysis is used routinely in mechanics, fluid mechanics and thermodynamics. In biology, however, the method has not received as much attention, even though it can provide quite some insight and explain many of the phenomena that we are used to see in daily life. In order to show this, let us consider, for example, the flight of birds. From observation we know that, in general, large birds fly faster than smaller ones. For simplicity, we assume that the bird's wings are shaped like the wings of an airplane. Then, the parameters that influence most the velocity of flight U are the mass of the bird m, the acceleration of gravity g, the area of the bird's wings A (i.e. the projection of the wings in the vertical direction), and the density of air ρ. Also, an important parameter is the angle of attack γ, measured as the angle between the horizontal and the chord of the wing. For simplicity, we restrict the analysis to geometrically similar birds and neglect the effects of viscosity and unsteadiness of the flow. Therefore, we seek a functional relation of the form

$$U = f(m, g, A, \rho, \gamma). \tag{3.88}$$

This gives rise to the dimension matrix

	U	m	g	A	ρ	γ
L	1	0	1	2	−3	0
M	0	1	0	0	1	0
T	−1	0	−2	0	0	0

The rank of this matrix is three and we obtain from dimensional analysis the result that three dimensionless products can be constructed. This results in

$$\frac{mg}{\rho AU^2} = f\left(\gamma, \frac{U^2}{g\sqrt{A}}\right). \tag{3.89}$$

The second dimensionless product is the Froude number, $\mathrm{Fr} = U^2/(g\sqrt{A})$. This quantity is important e.g. for free-surface flows. In the current situation we can neglect the influence of this dimensionless group, because we are in the high Froude number range.

Then, the expression given by Eq. (3.89) states that the weight of the bird mg and the aerodynamic lifting force ρAU^2 must be in balance and are a function of the angle of attack. In other words: The angle of attack must be adjusted in such a way that the aerodynamically produced lift carries the bird's weight. For geometrically similar birds $A \propto L^2$ and $L \propto m^{1/3}$ so that we have

$$A \propto m^{2/3}. \tag{3.90}$$

Then, we find from Eq. (3.89) for a constant angle of attack the following expression for the velocity

$$U \propto \left(\frac{m}{m^{2/3}}\right)^{1/2} \propto m^{1/6}. \tag{3.91}$$

This implies, that the characteristic flight velocity of birds increases with the weight and size of the bird. The increase in velocity is, however, slower than the increase in weight. Consider, for example, two birds with mass m_1 and m_2. Their velocities scale as

$$\frac{U_2}{U_1} = \left(\frac{m_2}{m_1}\right)^{1/6}. \tag{3.92}$$

If the second bird weighs twice as much as the first bird, it will fly only about 12% faster; if the second bird flies at twice the speed of the first bird, it will weigh 64-times as much as the first bird. From this small example we can see very nicely that from dimensional analysis and further physical reasoning and assumptions we obtain quite some insight in physics and can understand why things are like they are. However, one should keep in mind that we simplified the reality quite drastically, so that the ratios we are obtaining here for the flight speeds of two birds might be far from reality. So a goose has an average flight speed of about 90 km/h (average weight around 3.5 kg) whereas a swallow (average weight 0.022 kg) has one of 35 km/h. So taking the ratio of the weights, one would obtain a velocity ratio for the two birds of 7.4, whereas in reality it is around 2.5.

Another interesting result is found when we look at the frequency ω at which a bird flaps its wings. While flapping the wings, the wings move downward (or upward) with velocity $V \approx \omega L$, where L is the length of the wing, measured from tip to body. Therefore, we find

$$\frac{V}{U} = \frac{\omega L}{U}. \tag{3.93}$$

From aerodynamics we know that in ideal flow the ratio V/U is equal to the ratio of drag force F_D to lifting force F_L, therefore

$$\frac{V}{U} = \frac{F_D}{F_L} = \frac{c_D}{c_L} = \tan(\gamma), \tag{3.94}$$

where c_D and c_L are the drag and lift coefficients, respectively, defined by

$$c_D = \frac{F_D}{\rho A U^2} \quad \text{and} \quad c_L = \frac{F_L}{\rho A U^2}. \tag{3.95}$$

Both, c_D and c_L, are functions of γ, much like Eq. (3.89). It is reasonable to assume that birds adjust the angle of attack γ of their wings in such a way, that drag becomes minimum while lift becomes maximum, so that most of the bird's energy is used for forward flight and only little is consumed to overcome resistance. Then, $c_D/c_L = \text{const.}$, and this constant will be the same for all geometrically similar birds. It follows that $V/U = \text{const.}$, and with $L \propto m^{1/3}$ and Eq. (3.91) we find from Eq. (3.93)

$$\omega \propto m^{-1/6}. \tag{3.96}$$

This confirms the observation that small birds flap their wings faster than large ones (see Mc Mahon and Bonner 1985). It should be noted that a more refined description of bird's flight must take into account the unsteady movement of the wings, the viscosity of the air, and, of course, the fact that neither all birds are geometrically similar nor do they all move their wings in a similar fashion, so that the aerodynamics are more complicated. The reader is referred to Liebe and Liebe (2002), Liebe (2004) for more detailed investigations of these problems. Many other, most interesting examples of the application of dimensional analysis in biology are given by Stahl (1962) and Mc Mahon and Bonner (1985).

3.4.3 The Run-off from a Watershed

The watershed of a river is the territory that is drained by the river. If rain falls on a watershed, the river rises. The discharge of the river will continue to rise for a number of days after the rainfall, as some time is needed for the water to drain from the watershed into the river. Obviously, it is of practical importance to determine the amount of discharge $\dot{V}$ of the river as a function of the time t that has elapsed after the rainfall. Since the run-off $\dot{V}$ depends on the topography, the vegetation, the nature and saturation of the soil etc., we will restrict the analysis here to geologically and geometrically similar watersheds. In this case, the discharge $\dot{V}$ depends on time t, on the amount of rainfall H, which is usually expressed as "meter of rain" (i.e. the total rainfall of the watershed (in L^3) divided by the area of the watershed (in L^2)), on the area A of the watershed, on the acceleration of gravity g, on the density of the water ρ and on its kinematic viscosity ν. This leads to an expression of the form

$$f(\dot{V}, t, \nu, H, A, g, \rho) = 0. \tag{3.97}$$

As ρ is the only quantity that contains the dimension of mass, it is impossible to form a dimensionless product containing ρ. Consequently, ρ cannot enter into the problem. Then, we have $n = 6$ physical variables. The rank of the dimension matrix is $r = 2$, and we anticipate to find $d = n - r = 4$ dimensionless products. These are

$$\prod_1 = \dot{V}A^{-5/4}g^{-1/2}, \quad \prod_2 = tA^{-1/4}g^{1/2},$$
$$\prod_3 = \nu A^{-3/4}g^{-1/2} \quad \text{and} \quad \prod_4 = HA^{-1/2}. \tag{3.98}$$

Thus, we may express Eq. (3.97) in the equivalent dimensionless form

$$\frac{\dot{V}}{A^{5/4}g^{1/2}} = f\left(\frac{tg^{1/2}}{A^{1/4}}, \frac{\nu}{A^{3/4}g^{1/2}}, \frac{H}{A^{1/2}}\right). \tag{3.99}$$

It is reasonable to assume that the discharge $\dot{V}$ is proportional to the amount of rainfall H, so that $\dot{V}$ and H appear only in the combination $\dot{V}/H$. Thus, instead of $\prod_1$ we introduce

$$\prod_1^* = \prod_1 \prod_4^{-1} = \dot{V}A^{-3/4}g^{-1/2}H^{-1}, \tag{3.100}$$

which leads to the expression

$$\frac{\dot{V}}{A^{3/4}g^{1/2}H} = f\left(\frac{tg^{1/2}}{A^{1/4}}, \frac{\nu}{A^{3/4}g^{1/2}}\right). \tag{3.101}$$

This example was given by Langhaar (1951) together with experimental data. For different watersheds, the data of non-dimensional discharge as a function of non-dimensional time falls on a single curve, suggesting that the parameter $\Pi_3 = \nu A^{-3/4}g^{-1/2}$ has little influence on this problem. The discharge increases with time until it reaches a maximum for $tA^{-1/4}g^{1/2} \approx 3$, then it decreases. If one considers the complexity of the problem, it is astonishing to see how much insight and information can be gained without writing down or even solving the governing equations, but using purely dimensional reasoning.

3.5 Examples from Classical Electrodynamics

Classical electrodynamics covers the entire phenomenology of electricity and magnetism taking place on scales large enough to neglect quantum effects. The theory comprises quite a wide spectrum of sub-branches from electro- and magneto statics to electromagnetic waves, which however all condense down to a set of few equations that are briefly reviewed in the following, before diving into dimensional analytical aspects of electromagnetism.

3.5.1 *The Governing Equations*

The Maxwell equations are a set of partial differential equations that relate the behavior of electric and magnetic fields. In combination with the Lorentz force law the set of equations covers all phenomena of classical electrodynamics. The Maxwell equations relate the electric and magnetic fields **E** and **B** to each other, and to their sources. Generally the occurrence of electromagnetic fields is caused by electric charges and electric currents that are expressed here by the volumetric charge density $\rho = dq/dV$ and the current density $\mathbf{J} = d\mathbf{I}/dA$. In the SI-unit-system the macroscopic Maxwell equations read

$$\nabla \cdot \mathbf{D} = \rho, \tag{3.102}$$

$$\nabla \cdot \mathbf{B} = 0, \tag{3.103}$$

$$\nabla \times \mathbf{E} = -\frac{\partial \mathbf{B}}{\partial t}. \tag{3.104}$$

$$\nabla \times \mathbf{H} = \mathbf{J} + \frac{\partial \mathbf{D}}{\partial t}. \tag{3.105}$$

Here, **D** denotes the electric displacement field and **H** the magnetizing field. For linear isotropic materials the quantities are linearly connected by

$$\mathbf{D} = \varepsilon \mathbf{E}, \quad \mathbf{B} = \mu \mathbf{H}, \quad \mathbf{J} = \sigma \mathbf{E}. \tag{3.106}$$

The so called permittivity $\varepsilon = \varepsilon_r \varepsilon_0$ is a measure of the ability of a material to resist an electric field, thus how much electric field is generated per charge. For homogeneous materials, it is typically specified in terms of the relative permittivity ε_r relating it to the vacuum permittivity ε_0 in matter free space. In an analogous way $\mu = \mu_r \mu_0$ denotes the permeability as a measure of the capability of the material to be magnetized by an applied magnetic field. Again, it is characterized relatively to the permeability of absolute vacuum. The quantity σ characterizes the electric conductivity, thus the ability to conduct an electric current. The constants ε_0 and μ_0

are also referred to as electronic and magnetic constants and as shown in the next section are introduced as results of the chosen SI-unit-system.

Regarding the set of equations, Gauss's law in Eq. (3.102) relates the presence of an electric field around a charge, while its magnetic counterpart in Eq. (3.103) implies that the magnetic field is solenoidal, thus, there are no magnetic monopoles. Faraday's induction law in Eq. (3.104) relates an unsteady magnetic flux to the occurrence of an electric vortex field. The fourth equation, Ampère's circuit law, Eq. (3.105), describes that electric currents as well as changes in the electric field lead to a magnetic vortex field. While, the interaction between electric and magnetic fields is covered by the given set, the acting forces are subject of the so called Lorentz force law. A point charge q that moves through an electric or magnetic field with the velocity $\mathbf{v}$ experiences the force

$$\mathbf{F} = q(\mathbf{E} + \mathbf{v} \times \mathbf{B}). \tag{3.107}$$

As mentioned before the application of Maxwell's equations in combination with the Lorentz force law allows the derivation of the full phenomenology of classical electrodynamics. As a simple example, one might regard the electrostatic interaction between two electrically charged particles, where the first particle of charge q_1 is of distance r to the second particle of charge q_2. By choosing one of the particles as the source of the electric field, applying the divergence theorem for integrating Eq. (3.102) yields the acting electric field. The resulting force on the second test particle immediately follows the well-known Coulomb law

$$|\mathbf{F}| = \frac{q_1 q_2}{4\pi\varepsilon_0 r^2}. \tag{3.108}$$

The magneto static counterpart, at which two straight parallel conductors of constant currents I_1 and I_2 induce a magnetic field in their ambient and therefore cause a force upon each other. Ampère's force law can be derived from Maxwell's equations by integrating Eq. (3.105) and applying Lorentz law yielding the force per conductor unit length l

$$\frac{|\mathbf{F}|}{l} = \frac{\mu_0}{2\pi}\frac{I_1 I_2}{r}. \tag{3.109}$$

3.5.2 Electrodynamic Quantities in the SI-Unit-System

For defining an electrical base unit in principle one might apply different quantities, for instance the electric charge, the potential difference and many more. In the SI-system the electric current is introduced as defining quantity, while the remaining electromagnetic quantities are successively derived from the governing equations as presented in the following.

The unit Ampère is defined as the constant current flowing in two infinitely long thin wires, arranged in parallel to each other with a distance of 1 m, such that in vacuum they exert a magnetic force of exactly 2×10^{-7} N/m upon each other.

Based on this definition all remaining electric and magnetic units are derived, where only the most common are given here. The unit of the electric unit charge q, Coulomb, is defined as the charge that is transported by a constant current of one Ampère in one second, thus

$$\text{Coulomb} = \text{C} = \text{A s}. \tag{3.110}$$

The electric field $\mathbf{E}$ is defined as the electromagnetic force, i.e. the Lorentz force that it exerts on a stationary test particle of unit charge. The unit of the electric field thus is Newton per Coulomb. The electric potential ϕ, that is related to an electric field by $\mathbf{E} = -\nabla\phi$, denotes the electric potential a test particle of unit charge would have at any point within the electric field. Of course, this is equal to the work necessary to bring the test particle from infinity, where it has zero potential, to that point. The dimension of the potential is

$$\text{Volt} = \text{V} = \text{N m}/\text{C} = \text{kg m}^2\,\text{s}^{-3}\,\text{A}^{-2}. \tag{3.111}$$

In the context of electrical machines and generators the so called electromotive force defines the electric potential of the device in the circuit and accordingly carries the same dimension as ϕ. The capacitance C of a capacitor describes its ability to store electric charge. It is defined as the incremental change of charge per potential $C = dQ/d\phi$ and the unit is

$$\text{Farad} = \text{F} = \text{C}/\text{V} = \text{s}^4\,\text{A}^2\,\text{m}^{-2}\,\text{kg}^{-1}. \tag{3.112}$$

The electric resistance is a measure of the current over a conductor experiencing a potential difference of 1 V. The unit immediately follows from Ohm's law to

$$\text{Ohm} = \Omega = \text{V}/\text{A} = \text{kg m}^2\,\text{s}^{-3}\,\text{A}^{-2}. \tag{3.113}$$

The reciprocal of the resistance, the electric admittance, has the unit Siemens.

The magnetic field $\mathbf{B}$ is often defined by the force it exerts on a moving test particle, while the force is again determined via the Lorentz force law and is measured in Tesla

$$\text{Tesla} = \text{T} = \text{N s}/(\text{C m}) = \text{kg s}^{-2}\,\text{A}^{-1}. \tag{3.114}$$

The corresponding magnetic surface flux is therefore given in

$$\text{Weber} = \text{Wb} = \text{T m}^2 = \text{kg m}^2\,\text{s}^{-2}\,\text{A}^{-1}. \tag{3.115}$$

An electric current flowing through a circuit causes a proportional magnetic field around it. The change of the electric flux causes a temporal change of the magnetic field, which then induces an electromotive force in the circuit. Now the so called self-inductance L denotes the property that a change in current induces a proportional electromotive force in the conductor of LdI/dt. Of course, the varying field in this circuit also induces an electromotive force in any adjacent circuit and the proportionality constant then termed mutual inductance coefficient. In the SI system, the unit of the inductance is the

$$\text{Henry} = \text{H} = \text{V s}/\text{A} = \text{kg m}^2\,\text{s}^{-2}\,\text{A}^{-2}. \tag{3.116}$$

3.5.3 *Implication of the SI-Unit-System*

In the classical literature on electrodynamics, one often finds the use of alternate unit systems, for instance the Gaussian system. The mechanical units in the SI and Gaussian system are based on the same quantities, however with different prefixes. As one of several unit systems based on the CGS (centimeter, gram, second) system in the Gaussian system, centimeters are used instead of meters, grams instead of kilograms, while the base unit for time is the same in both systems. That means there exists a one to one correspondence, i.e. bijection, between the two unit systems and in consequence a corresponding accordance of all the derived quantities. For instance the unit for force, i.e. the dyne is easily converted into its SI equivalent by

$$1\,\text{dyne} = 1\,\text{g cm s}^{-2} = 10^{-5}\,\text{N}. \tag{3.117}$$

Thus, we see that as discussed more generally in Sect. 1.2.4, the conversion from one base system to the other, only necessitates an according conversion factor that immediately follows from the prefixes. However, in case of the electromagnetic units, the situation becomes more complex. Other than in the SI-system where Ampère is introduced as electrical base unit, in the Gaussian system no new electric unit is added. Instead, all quantities are based on the mechanical base units for mass, length and time. The unit charge is defined such that two unit charges, placed in vacuum, separated by a distance of 1 cm, exert a force of one dyne onto each other. With Coulomb's law where $F \sim q_1 q_2 / r^2$ the unit for the electric charge immediately follows to $\text{N}^{1/2}$ m or $\text{kg}^{1/2}\,\text{m}^{3/2}\,\text{s}^{-1}$. In consequence, the electric current then carries the dimension $\text{kg}^{1/2}\,\text{m}^{3/2}$.

According to our previous discussion in Sect. 1.4, including an additional base variable results into an additional proportionality constant. Therefore, in case of the SI-unit-system, the introduction of Ampère as base unit must result in the appearance of an additional constant, purely reasoned by dimensional considerations. For Coulomb's law to be dimensionally homogeneous the proportionality constant enters, yielding $F \sim q_1 q_2 / r^2$. Keeping in mind the relation between a unit charge and the electric current, the constant carries the dimension N m^2 s^{-2} A^{-2} or kg m^3 s^{-4} A^{-2}. In practice the inverse of this constant is applied and is denoted as the introduced electric constant ε_0. Analogously to the given argumentation in terms of Coulomb's force law one might also argue by Ampère's force law for currents through two straight parallel conductors, which is associated with the introduction of the magnetic constant μ_0 as proportionality constant. However, both paths are equivalent showing that the introduction of an electromagnetic unit results in the occurrence of proportionality constant.

The values of the constants follow from the definition of the electromagnetic unit. But before establishing the values it is essential to consider that the electric and magnetic constants ε_0 and μ_0 are interconnected by the speed of light. Regarding Maxwell's equations for a region with no charges and no currents, as in vacuum, and applying the curl of Eqs. (3.104) and (3.105) in combination with the solenoidal fields in Eqs. (3.102) and (3.103), yields the wave equations

$$\frac{\partial^2 \mathbf{E}}{\partial t^2} - \frac{1}{\varepsilon_0 \mu_0} \Delta \mathbf{E} = 0, \quad \frac{\partial^2 \mathbf{B}}{\partial t^2} - \frac{1}{\varepsilon_0 \mu_0} \Delta \mathrm{B} = 0 \tag{3.118}$$

for the electric and the magnetic field. The wave velocity in both cases corresponds to the speed of light resulting in

$$c_0 = \frac{1}{\sqrt{\varepsilon_0 \mu_0}}. \tag{3.119}$$

Now the constants are set directly by applying the given definition of the Ampère to Ampère's force law, yielding the magnetic constant to

$$\mu_0 = 4\pi \times 10^{-7} \ \mathrm{N/A^2} \tag{3.120}$$

and consequently the electric constant to

$$\varepsilon_0 = \frac{1}{\mu_0 c_0^2} \approx 8.854 \times 10^{-12} \ \mathrm{F/m}. \tag{3.121}$$

The definition of the SI-unit-system is well suited to reflect how the choice of the base system affects the associated physical formulation. Introducing an additional unit incorporates an extra proportionality constant necessary for dimensional homogeneity of the governing equations. In the present case, there are even two constants occurring, that however are interrelated so that one of them has to be defined. This definition is somewhat arbitrary chosen by the way the new unit is defined.

Analogously to the previous reasoning given in Sect. 1.4 neglecting the vacuum permittivity in dimension analysis automatically implies that the validity of Coulombs law is neglected. This assumption is typically justified, when regarding problems that are not dominated by the force interaction of few charges, but that involve a large number of charges, as for instance in the case of electrical circuits.

3.5.4 Electrical Circuits

In an electric circuit, an inductor (L), a capacitator (C) and a resistor (R) are connected in series or in parallel as shown in Fig. 3.12. In practice they are typically used in electric devices such as radio receivers or TV sets for tuning signals to a particular frequency or as band pass filters. For some given configuration, we want to investigate the characteristic time for the circuit configuration, thus

$$\tau = f(L, C, R). \tag{3.122}$$

The dimension matrix then reads

	τ	L	C	R
L	0	2	−2	2
M	0	1	−1	1
T	1	−2	4	−3
I	0	−2	2	−2

The number of variables is obviously $n = 4$, but the matrix does not have the full rank. Regarding the [LMI] exponents of the quantities L, C and R one immediately sees that they are multiples of each other. Thus, it is found that the rank is $r = 2$ and we therefore obtain two non-dimensional products ($d = 2$) to

$$\frac{\tau}{\sqrt{LC}} = f\left(\frac{R^2 C}{L}\right). \tag{3.123}$$

This relation includes quite some useful information. The first dimensionless product characterizes a circuit of negligible resistance, thus a pure LC-circuit, for which one assumes that no energy is dissipated, yielding a harmonic oscillator.

Fig. 3.12 *LCR* series and parallel circuits

With regard to the non-dimensional coefficients the second product is neglected and the natural oscillation period follows to

$$\frac{\tau}{\sqrt{LC}} = \text{const.} \tag{3.124}$$

The second dimensionless product determines the damping, which controls the type of transient the circuit exhibits, i.e. whether the circuit response is underdamped, critically damped or overdamped. As another example one might regard the special case of an *RC*-circuit that consists of a charged capacitor and a resistor. The capacitor discharges the energy through the resistor so that by means of dimensional analysis the products in Eq. (3.123) reduce to a single product. Rearranging the dimensionless products in Eq. (3.123) to eliminate L yields the characteristic discharge time

$$\frac{\tau}{RC} = \text{const.} \tag{3.125}$$

In order to obtain the precise transient behavior it is necessary to apply the according modelling equations, e.g. Kirchhoff's circuit laws, to the regarded circuit topology and subsequently solve the resulting differential equations. So for instance for a parallel *RC*-circuit the voltage decay is exponential, while for a series circuit it follows a rational functionality. However, the obtained non-dimensional product and therefore the characteristic discharge time remains, regardless of the exact setup.

Chapter 4
Similarity Solutions

As already mentioned, dimensional analytical considerations provide exact results. The solution of a problem is considerably simplified, however in most cases the result must be complemented by experimental evidence or by solving the differential equations describing the problem.

In general, physical problems are described by sets of partial differential equations, which often can only be solved numerically with tremendous effort. However, in some cases, the governing partial differential equations can be reduced to ordinary differential equations that are more easily accessible. If we succeed, we often speak of "exact" solutions (in particular in relation to the Navier-Stokes equations), even if these ordinary differential equations need to be integrated numerically.

Within this group of equations a particular class are the so-called "similarity solutions". The name arises from the fact that the independent variables for these problems can be combined to so called "similarity variables", so that the problem depends only on the similarity variable, leading to the reduction of the partial differential equations to ordinary differential equations. In the development of such solutions dimensionless analytical approaches play an important role, so that selected examples are presented in the following.

4.1 Transient Temperature Distribution in a Semi-infinite Body

We consider a semi-infinite body (Fig. 4.1), that is bounded at $x = 0$ infinitely extends in x- direction. It is subjected to the temperature $\vartheta = \vartheta_0$ at time $t = 0$ for $x \geq 0$. At time $t \geq 0$ the body has the temperature $\vartheta = \vartheta_\infty$ on its surface, and we search for the temporal and the spatial variation of the temperature $\vartheta(t, x)$.

V. Simon et al., *Dimensional Analysis for Engineers*,
Mathematical Engineering, DOI 10.1007/978-3-319-52028-5_4

Mathematically, the problem is described by the one-dimensional energy equation for an unsteady process

$$\frac{\partial \vartheta}{\partial t} = \alpha \frac{\partial^2 \vartheta}{\partial x^2}, \quad \alpha = \frac{k}{\rho c}, \tag{4.1}$$

with the boundary conditions

$$t = 0, \quad x \geq 0: \quad \vartheta = \vartheta_0 \tag{4.2}$$

$$t \geq 0, \quad x = 0: \quad \vartheta = \vartheta_\infty. \tag{4.3}$$

The solution for the described problem must have the form

$$\vartheta = f(t, x, \alpha, \vartheta_0, \vartheta_\infty). \tag{4.4}$$

Due to the realization that in a heat transfer problem only temperature differences are significant, Eq. (4.4) can also be written in the form

$$\vartheta - \vartheta_0 = f(t, x, \alpha, \vartheta_\infty - \vartheta_0), \tag{4.5}$$

so that the dimension matrix is obtained as

	$\vartheta - \vartheta_0$	$\vartheta_\infty - \vartheta_0$	α	t	x
L	0	0	2	0	1
M	0	0	0	0	0
T	0	0	−1	1	0
Θ	1	1	0	0	0

Rearranging yields

	$(\vartheta - \vartheta_0)/(\vartheta_\infty - \vartheta_0)$	$\vartheta_\infty - \vartheta_0$	$x/\sqrt{\alpha t}$	t	x
L	0	0	0	0	1
M	0	0	0	0	0
T	0	0	0	1	0
Θ	0	1	0	0	0

and thus one obtains

$$\frac{\vartheta - \vartheta_0}{\vartheta_\infty - \vartheta_0} = f\left(\frac{x}{\sqrt{\alpha t}}\right). \tag{4.6}$$

Thus, the independent variables occur only in the combination $\eta = x/\sqrt{\alpha t}$, and η is referred to as similarity variable. Thus, the number of independent variables in the mathematical sense is reduced from two to one. The reason to understand, why this happens in the case under investigation is relatively easy. We notice that no length scale is present in this problem and thus the coordinate x must be made dimensionless by a suitable choice of the other quantities. Inserting the expressions

$$\frac{\vartheta - \vartheta_0}{\vartheta_\infty - \vartheta_0} = F(\eta) \quad \text{and} \quad \eta = \frac{x}{\sqrt{4\alpha t}} \tag{4.7}$$

into Eq. (4.1), the following ordinary differential equation will be obtained

$$F'' + 2\eta F' = 0, \tag{4.8}$$

with the boundary conditions

$$\eta = 0: \quad F(\eta = 0) = 1 \tag{4.9}$$

and

$$\eta \to \infty: \quad F(\eta \to \infty) = 0. \tag{4.10}$$

The solution of Eq. (4.8) is obtained by using the substitution $z = F'$ and then integrating twice. One obtains

$$F(\eta) = C_1 \int_0^\eta \exp(-\eta^2)\mathrm{d}\eta + C_2, \tag{4.11}$$

where the constants are determined by the boundary conditions given by Eqs. (4.9) and (4.10) to

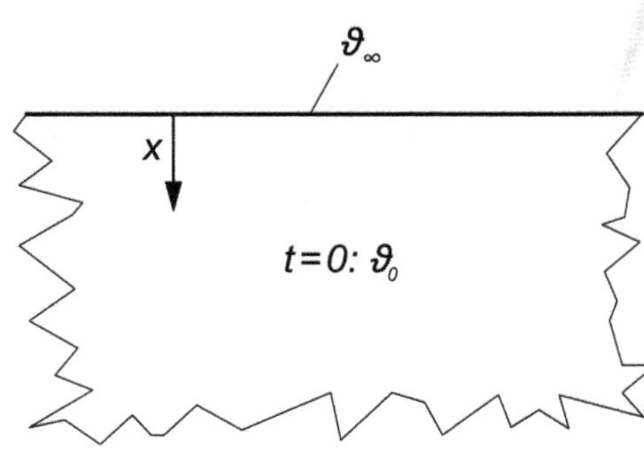

Fig. 4.1 Half-infinite body

$$C_2 = 1 \quad \text{and} \quad \frac{1}{C_1} = -\int_0^\infty \exp(-\eta^2)d\eta = -\frac{\sqrt{\pi}}{2}. \tag{4.12}$$

We thus obtain the solution of Eq. (4.8) to be

$$F(\eta) = 1 - \text{erf}(\eta), \tag{4.13}$$

whereas

$$\text{erf}(\eta) = \frac{2}{\sqrt{\pi}} \int_0^\eta \exp(-\eta^2)d\eta \tag{4.14}$$

is the error function (see Bronstein and Semendjajew 1981)

4.2 Flow Along a Stretched Sheet

In the following example we consider the two-dimensional flow which is set up when a sheet is stretched with the constant strain rate a at the plane $x = 0$, so that the sheet speed is $u(x) = ax$, as shown in Fig. 4.2. Due to the no-slip condition, the surrounding fluid is entrained, and the resulting steady flow is governed by the continuity equation

$$\frac{\partial u}{\partial x} + \frac{\partial v}{\partial y} = 0, \tag{4.15}$$

and the x- and y-components of the momentum balance

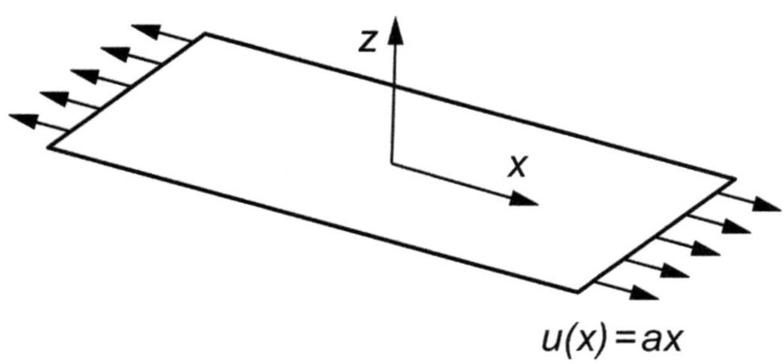

Fig. 4.2 Stretched sheet

$$u\frac{\partial u}{\partial x}+v\frac{\partial u}{\partial y}=-\frac{1}{\rho}\frac{\partial p}{\partial x}+\nu\left(\frac{\partial^2 u}{\partial x^2}+\frac{\partial^2 u}{\partial y^2}\right), \quad (4.16)$$

$$u\frac{\partial v}{\partial x}+v\frac{\partial v}{\partial y}=-\frac{1}{\rho}\frac{\partial p}{\partial y}+\nu\left(\frac{\partial^2 v}{\partial x^2}+\frac{\partial^2 v}{\partial y^2}\right), \quad (4.17)$$

with the boundary conditions

$$y=0: \quad u=ax \quad \text{and} \quad v=0 \quad (4.18)$$

$$y\to\infty: \quad u=0 \quad \text{and} \quad p=p_0. \quad (4.19)$$

The dimension matrix for this case then becomes

	u	a	ν	x	y
L	1	0	2	1	1
M	0	0	0	0	0
T	−1	−1	−1	0	0

respectively

	$u/(ax)$	a	$\sqrt{\nu/a}$	x	y
L	0	0	1	1	1
M	0	0	0	0	0
T	0	−1	0	0	0

yielding

$$\frac{u}{ax}=U\left(x\sqrt{a/\nu},y\sqrt{a/\nu}\right). \quad (4.20)$$

Carrying out corresponding considerations for the velocity component v and the pressure p we obtain

$$\frac{v}{ay}=V\left(x\sqrt{a/\nu},y\sqrt{a/\nu}\right) \quad (4.21)$$

and

$$\frac{p_0-p}{\rho a\nu}=P\left(x\sqrt{a/\nu},y\sqrt{a/\nu}\right). \quad (4.22)$$

Of course, from a purely dimension analytical point of view one could also choose the dimensionless products $u/(ay)$ or $v/(ax)$, however the products in Eqs. (4.21) and (4.22) are well suited with regard to the boundary conditions given by Eq. (4.18).

In a mathematical sense, no reduction in the number of variables has occurred. A further simplification is however possible, if one considers that the boundary conditions given by Eqs. (4.18) and (4.19) can also be satisfied if the dependence of x in the functions U, V and P disappears. With

$$\eta = y\sqrt{\frac{a}{\nu}}, \quad \frac{u}{ax} = U(\eta), \tag{4.23}$$

$$\frac{v}{ay} = V(\eta) \quad \text{and} \quad \frac{p_0 - p}{\rho a \nu} = P(\eta)$$

as well as the continuity Eq. (4.15) one obtains the relation

$$U + V + \eta V' = 0 \quad \text{respectively} \quad U = -(\eta V)' = F'(\eta), \tag{4.24}$$

and thus the functions are assumed to be

$$u = axF' \quad \text{and} \quad v = -\sqrt{a\nu}F. \tag{4.25}$$

The same result is achieved by introducing the stream-function

$$\psi = \sqrt{a\nu}xF \tag{4.26}$$

into Eq. (4.16) which yields the ordinary differential equation

$$F'^2 - FF'' = F'''. \tag{4.27}$$

From Eq. (4.17) the equation

$$P' = FF' + F'' \quad \text{respectively} \quad P = \frac{1}{2}F^2 + F' + C \tag{4.28}$$

can be obtained. The boundary conditions are given by

$$\eta = 0: \quad F = 0, \quad F' = 1, \tag{4.29}$$

$$\eta \to \infty: \quad F' = 0, \quad \text{and} \quad P = 0. \tag{4.30}$$

The solution of (4.27), that fulfills the boundary conditions given in Eqs. (4.29) and (4.30) was obtained by Crane (1970) as

$$F = 1 - \exp(-\eta), \tag{4.31}$$

so that with Eqs. (4.25) and (4.28) for the velocity components and the pressure relations one finally obtains

$$u = ax \exp(-y\sqrt{a/v}), \tag{4.32}$$

$$\mathrm{v} = -\sqrt{a\mathrm{v}}(1 - \exp(-\eta)) \tag{4.33}$$

and

$$\frac{p_0 - p}{\rho} = a\mathrm{v}\left(\frac{1}{2}(1 - \exp(-\eta))^2 - \exp(-\eta)\right). \tag{4.34}$$

4.3 Natural Convection Along a Vertical Plate

Next, we consider a hot, vertical plate, whose temperature ϑ_w is higher than the ambient air temperature ϑ_0. In the vicinity of the plate the ambient air heats up and flows upward opposed to the gravity field, as the density decreases with increasing temperature. This is visualized in Fig. 4.3. Under the Boussinesq approximation, the Navier-Stokes equations take the form (Kays et al. 2004)

$$\frac{D\vec{u}}{Dt} = -\frac{1}{\rho_0}\nabla p + v\nabla^2\vec{u} - \vec{g}\beta\Delta\vartheta, \tag{4.35}$$

where

$$\beta = -\frac{1}{\rho_0}\left(\frac{\partial\rho}{\partial\vartheta}\right)_p \tag{4.36}$$

is the thermal expansion coefficient, which becomes $\beta = 1/\vartheta_0$ for ideal gas, Kays et al. (2004). Applying x vertically upward along the plate and y perpendicular to the plate, we obtain the continuity equation in the form

$$\frac{\partial u}{\partial x} + \frac{\partial v}{\partial y} = 0, \tag{4.37}$$

and the components of the momentum balance in x- and y-direction result to be after using the boundary layer simplifications for this free convection flow (Kays et al. 2004; Eckert and Drake 1972)

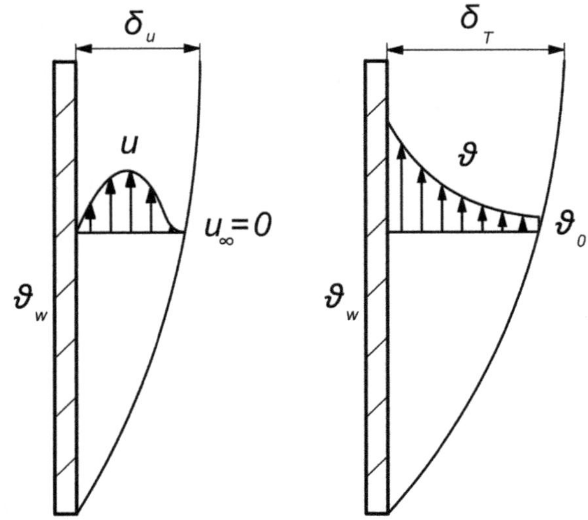

Fig. 4.3 Velocity and temperature field for a free convection flow along a vertical plate

$$u\frac{\partial u}{\partial x}+v\frac{\partial u}{\partial y}=-\frac{1}{\rho_0}\frac{\partial p}{\partial x}+g\beta(\vartheta-\vartheta_0)+\nu\frac{\partial^2 u}{\partial y^2} \tag{4.38}$$

and

$$0=\frac{\partial p}{\partial y}. \tag{4.39}$$

The energy equation takes the following form after invoking the boundary layer simplifications (Kays et al. 2004)

$$u\frac{\partial \vartheta}{\partial x}+v\frac{\partial \vartheta}{\partial y}=\alpha\frac{\partial^2 \vartheta}{\partial y^2}. \tag{4.40}$$

Because of Eq. (4.39) it follows that $p=p(x)$, so that Eq. (4.38) for large y, i.e. outside the boundary layer gives the result

$$\frac{\mathrm{d}p}{\mathrm{d}x}=-\rho_0 g\beta\Delta\vartheta=0. \tag{4.41}$$

This yields Eq. (4.38) in the form

$$u\frac{\partial u}{\partial x}+v\frac{\partial u}{\partial y}=g\beta(\vartheta-\vartheta_0)+\nu\frac{\partial^2 u}{\partial y^2}. \tag{4.42}$$

We therefore expect a functional dependence for u and ϑ of the form

$$u = f(x, y, \nu, \alpha, g\beta, \Delta\vartheta) \tag{4.43}$$

and

$$\vartheta - \vartheta_w = f(x, y, \nu, \alpha, g\beta, \Delta\vartheta) \tag{4.44}$$

with $\Delta\vartheta = \vartheta_w - \vartheta_0$. Again the dimensionless products may be found by evaluating the dimension matrix. Instead an alternate way is shown to obtain the relevant dimensionless quantities that are applicable if the equations describing the problem are explicitly given. For this purpose, the magnitudes of the individual terms in the differential equations are estimated and it is requested that the individual terms are of the same order of magnitude. If one chooses L and δ as typical length scales in x- and y-direction, then the magnitudes of the individual terms follow to

$$u\frac{\partial u}{\partial x} \approx \frac{U^2}{L} \tag{4.45}$$

and

$$\nu\frac{\partial^2 u}{\partial y^2} \approx \frac{\nu U}{\delta^2}. \tag{4.46}$$

Because the convective terms must be of the same order of magnitude as the buoyancy term, one obtains the estimate

$$U \approx \sqrt{Lg\beta\Delta\vartheta}, \tag{4.47}$$

and from the requirement that the convective and viscous terms in Eq. (4.42) also must be of the same order, one gets

$$\delta \approx \sqrt{\frac{\nu L}{U}} \quad \text{respectively} \quad \delta \approx \left(\frac{\nu^2 L}{g\beta\Delta\vartheta}\right)^{1/4}. \tag{4.48}$$

This allows to write Eqs. (4.43) and (4.44) to be written in the form

$$u = \sqrt{Lg\beta\Delta\vartheta}\; f\left(\frac{x}{L}, y\left(\frac{g\beta\Delta\vartheta}{\nu^2 L}\right)^{1/4}, \frac{\nu}{\alpha}\right) \tag{4.49}$$

and

$$\frac{\vartheta - \vartheta_0}{\vartheta_w - \vartheta_0} = f\left(\frac{x}{L}, y\left(\frac{g\beta\Delta\vartheta}{\nu^2 L}\right)^{1/4}, \frac{\nu}{\alpha}\right). \tag{4.50}$$

Since the presently considered problem of an infinitely long plate does not imply a geometrical length L, we have to choose $L = \mathrm{x}$, so that both the velocity field and the temperature only depend on the similarity variable

$$\eta = y\left(\frac{g\beta\Delta\vartheta}{4\nu^2 \mathrm{x}}\right)^{1/4} \tag{4.51}$$

and the Prandtl number

$$\mathrm{Pr} = \frac{\nu}{\alpha}. \tag{4.52}$$

In terms of the stream-function ψ it thus follows

$$u = \frac{\partial\psi}{\partial y} \quad \text{and} \quad v = -\frac{\partial\psi}{\partial x}. \tag{4.53}$$

which is of the order of magnitude $\psi \approx U\delta$, thus

$$\psi = \left(x^3\nu^2 g\beta\Delta\vartheta/4\right)^{1/4} F(\eta, \mathrm{Pr}). \tag{4.54}$$

For the temperature field we assume

$$\frac{\vartheta - \vartheta_0}{\vartheta_w - \vartheta_0} = \theta(\eta, \mathrm{Pr}), \tag{4.55}$$

and the differential Eqs. (4.40) and (4.42) simplify to

$$F''' + 3FF'' - 2F' + \theta = 0 \tag{4.56}$$

and

$$\theta'' + 3\,\mathrm{Pr}\,F\theta' = 0 \tag{4.57}$$

with the boundary conditions

$$\eta = 0: \quad F = 0, \quad F' = 0 \quad \theta = 1 \tag{4.58}$$

$$\eta \to \infty: \quad F' = 0, \quad \theta = 0. \tag{4.59}$$

Equations (4.56) and (4.57) have to be solved numerically. This however is no problem, because these equations are just ordinary differential equations.

4.4 Natural Convection Along a Horizontal Cylinder

As another example for natural convection, we consider a horizontal cylinder with a radius R having the constant temperature ϑ_w and is surrounded by liquid of temperature ϑ_0. For $\vartheta_w > \vartheta_0$ the liquid is heated in the vicinity of the cylinder and flows within a thin boundary layer against the direction of gravity towards the top of the cylinder (see Fig. 4.4). The origin of the x-coordinate with $x = 0$ is set at the lower end of the cylinder and is directed along the cylinder circumference, while the y-coordinate is chosen perpendicular to the cylinder surface. The conservation equations for mass, momentum and energy read after applying the boundary layer simplifications and using the Boussinesq approximation (Kays et al. 2004)

$$\frac{\partial u}{\partial x} + \frac{\partial v}{\partial y} = 0, \tag{4.60}$$

$$u\frac{\partial u}{\partial x} + v\frac{\partial u}{\partial y} = g\beta(\vartheta - \vartheta_0)\sin\left(\frac{x}{R}\right) + \nu\frac{\partial^2 u}{\partial y^2}, \tag{4.61}$$

and

$$u\frac{\partial \vartheta}{\partial x} + v\frac{\partial \vartheta}{\partial y} = \alpha\frac{\partial^2 \vartheta}{\partial y^2}. \tag{4.62}$$

The magnitudes of the convection and buoyancy terms in Eq. (4.61) are

$$u\frac{\partial u}{\partial x} \approx \frac{U^2}{R} \tag{4.63}$$

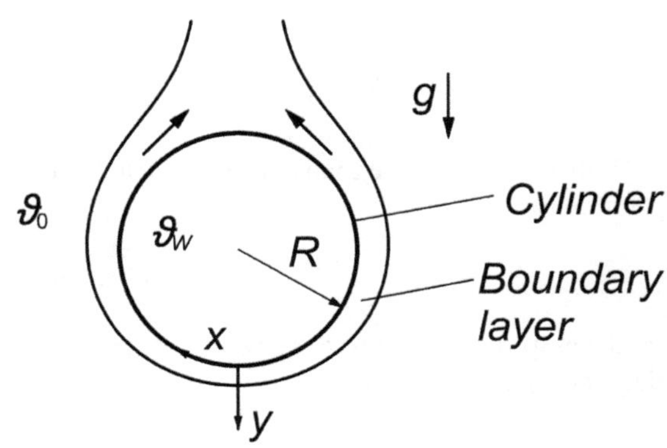

Fig. 4.4 Natural convection around a horizontal cylinder

and

$$\frac{U^2}{R} \approx g\beta\Delta\vartheta \quad \text{with} \quad \Delta\vartheta = \vartheta_w - \vartheta_0, \tag{4.64}$$

and thus as a measure of the flow velocity it follows that

$$U \approx \sqrt{Rg\beta\Delta\vartheta}. \tag{4.65}$$

Since the convection and friction terms must also be of the same order of magnitude it is found that

$$\frac{U^2}{R} \approx \nu \frac{U}{\delta^2}, \tag{4.66}$$

yielding the order of the boundary layer thickness to

$$\delta \approx \left(\frac{\nu^2 R}{g\beta\Delta\vartheta}\right)^{1/4}. \tag{4.67}$$

Because of $\psi \approx U\delta$ one obtains

$$\psi = \left(R^3\nu^2 g\beta\Delta\vartheta/4\right)^{1/4} \quad F(\xi,\eta) \tag{4.68}$$

and

$$\frac{\vartheta - \vartheta_0}{\vartheta_w - \vartheta_0} = \theta(\xi,\eta), \tag{4.69}$$

where

$$\eta = \frac{y}{\delta} = y\left(\frac{g\beta\Delta\vartheta}{4\nu^2 R}\right)^{1/4} \quad \text{and} \quad \xi = \frac{x}{R}. \tag{4.70}$$

With regard to the expansion of the sine function

$$\sin(\xi) = \xi - \frac{1}{3!}\xi^3 + \frac{1}{5!}\xi^5 - \cdots \tag{4.71}$$

we develop the functions $F(\xi,\eta)$ and $\theta(\xi,\eta)$ with regard to ξ (Crane, unpublished discussion) and thus set

$$F(\xi,\eta) = \xi F_0(\eta) + \xi^3 F_1(\eta) + \xi^5 F_2(\eta) + \cdots \tag{4.72}$$

and

$$\theta(\xi,\eta) = \theta_0(\eta) + \xi^2\theta_1(\eta) + \xi^4\theta_2(\eta) + \cdots \tag{4.73}$$

The velocity components are then given by Eq. (4.53) as

$$u = \sqrt{Rg\beta\Delta\vartheta}\quad \left(\xi F_0' + \xi^3 F_1' + \xi^5 F_2' + \cdots\right) \tag{4.74}$$

and

$$v = -\sqrt{g\beta\Delta\vartheta/R}\ \left(F_0 + 3\xi^2 F_1 + 5\xi^4 F_2 + \cdots\right). \tag{4.75}$$

The Eqs. (4.61) and (4.62) can be converted into a set of ordinary differential equations whose first terms result in

$$F_0''' = (F_0')^2 - F_0 F_0'' \tag{4.76}$$

and

$$\theta_0'' = -\,\mathrm{Pr}\ F_0\,\theta_0' \tag{4.77}$$

with the boundary conditions

$$\eta = 0 \quad F_i = 0,\ \ F_i' = 0,\ \ \theta_0 = 1,\ \ \theta_{i+1} = 0,\ \ i = 0, n \tag{4.78}$$

and

$$\eta \to \infty \quad F_i' = 0,\ \ \theta_i = 0,\ \ i = 0, n. \tag{4.79}$$

From a numerical integration of Eqs. (4.76) and (4.77) for Pr = 0.7 one obtains (Crane-unpublished discussion)

$$F_0(\eta \to \infty) \cong 1.0 \quad \text{and} \quad \theta_0'(0) \cong -0.42. \tag{4.80}$$

4.5 Laminar Free Jet

We now consider a liquid jet flowing from a circular orifice into a silent ambient of the same liquid (Fig. 4.5). Due to viscosity the ambient of the jet is also set in motion, so that with increasing distance from the orifice the jet width increases. Only in the immediate vicinity of the exit opening, the flow is affected by the geometry of the orifice, and at some distance from the orifice the conditions become independent from the geometrical form and the flow only remembers the discharge momentum. We assume the jet at the origin to be generated by a virtual momentum

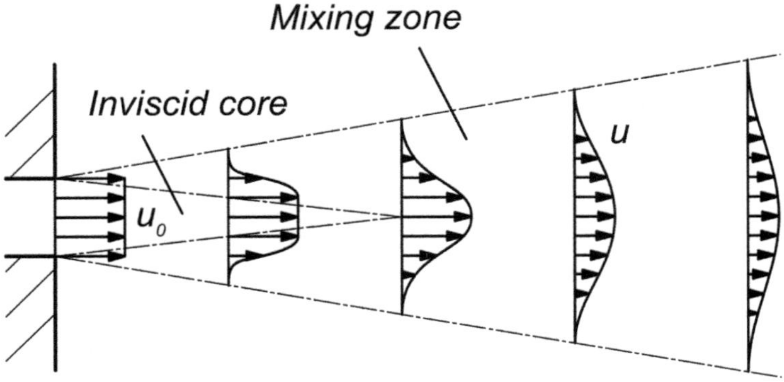

Fig. 4.5 Free jet

source, and apply the x-coordinate in the axial direction of the jet and the coordinate r perpendicular to the jet axis. The kinematic momentum of the jet is then given by Schlichting (1979)

$$K = \frac{J}{\rho} = 2\pi \int_0^\infty u^2 r \mathrm{d}r = \text{const.} \tag{4.81}$$

and has the same value for each axial position x. We assume that the temperature ϑ of the emerging jet takes a different value than the ambient temperature ϑ_∞ so that the kinematic enthalpy flux is given by

$$H = 2\pi \int_0^\infty ur(\vartheta - \vartheta_\infty) \mathrm{d}r = \text{const.} \tag{4.82}$$

The enthalpy flux is also independent of the axial position x. For the considered rotationally symmetric case, the continuity equation, the momentum balance in the axial direction and the heat balance read if we again use boundary layer simplifications (see Schlichting 1979)

$$\frac{\partial}{\partial x}(ur) + \frac{\partial}{\partial r}(vr) = 0, \tag{4.83}$$

$$u\frac{\partial u}{\partial x} + v\frac{\partial u}{\partial r} = \frac{\nu}{r}\frac{\partial}{\partial r}\left(r\frac{\partial u}{\partial r}\right) \tag{4.84}$$

and

$$u\frac{\partial\vartheta}{\partial x}+v\frac{\partial\vartheta}{\partial r}=\frac{v}{\mathrm{Pr}}\frac{1}{r}\frac{\partial}{\partial r}\left(r\frac{\partial\vartheta}{\partial r}\right). \tag{4.85}$$

The velocity and temperature fields are subjected to the boundary conditions

$$r=0: \quad v=0, \quad \frac{\partial u}{\partial r}=0, \quad \frac{\partial\vartheta}{\partial r}=0, \tag{4.86}$$

$$r\to\infty: \quad u\to 0, \quad \vartheta\to\vartheta_{\infty}. \tag{4.87}$$

The stream-function in the present rotationally symmetric case is defined by Eckert and Drake (1979)

$$ru=\frac{\partial\psi}{\partial r} \quad \text{and} \quad rv=-\frac{\partial\psi}{\partial x}. \tag{4.88}$$

To obtain suitable dimensionless products, we again estimate the magnitudes of the individual terms in the equation, where L is the characteristic length in the x-direction and the jet width R represents the characteristic length in the radial direction. We obtain the estimate for the inertia terms

$$u\frac{\partial u}{\partial x}\approx\frac{U^2}{L}, \tag{4.89}$$

and for the viscous term

$$\frac{v}{r}\frac{\partial}{\partial r}\left(r\frac{\partial \mathrm{u}}{\partial r}\right)\approx\frac{vU}{R^2}. \tag{4.90}$$

From the demand that both terms are of the same magnitude it follows that

$$\frac{U^2}{L}\approx\frac{vU}{R^2}. \tag{4.91}$$

From the momentum balance, given by Eq. (4.81), we find the estimate

$$U^2R^2\approx K. \tag{4.92}$$

Thus, the magnitudes of the axial velocity and radial extent of the jet are given by

$$U\approx\frac{K}{vL} \quad \text{and} \quad R\approx\frac{vL}{\sqrt{K}}. \tag{4.93}$$

Due to $\psi\approx UR^2$ we obtain a measure of the stream-function by

$$\psi \approx \nu L. \tag{4.94}$$

Thus, for the stream-function we can write

$$\psi = (\nu L)F\left(\frac{x}{L}, \frac{r}{R}\right) = (\nu L)F\left(\frac{x}{L}, \frac{r}{L}\frac{\sqrt{K}}{\nu}\right) \tag{4.95}$$

and, as for the length L no other characteristic value than x can be assigned, it follows that

$$\psi = \nu x F(\eta) \quad \text{with} \quad \eta = \frac{r}{x}\frac{\sqrt{cK}}{\nu}, \tag{4.96}$$

where the constant c has been inserted in order to satisfy the integral momentum balance. The velocity components follow from Eqs. (4.88) and (4.96) to

$$u = \frac{cK}{\nu x}\frac{F'}{\eta} \quad \text{and} \quad v = \frac{\sqrt{cK}}{x}\left(F' - \frac{F}{\eta}\right), \tag{4.97}$$

so that the momentum balance given by Eq. (4.84) results in

$$\frac{FF'}{\eta^2} - \frac{F'^2}{\eta} - \frac{FF''}{\eta} = F''' + \frac{F'}{\eta^2} - \frac{F''}{\eta} \tag{4.98}$$

or

$$-\left(\frac{FF'}{\eta}\right)' = \left(F'' - \frac{F'}{\eta}\right)'. \tag{4.99}$$

The boundary conditions become

$$\eta = 0: \quad F = F' = F'' = 0. \tag{4.100}$$

Equation (4.99) can be integrated to give

$$\eta F'' = F' - FF', \tag{4.101}$$

where the integration constant for Eq. (4.100) is determined to be zero. Equation (4.101) can be cast into the following form

$$(\eta F')' = 2F' - \frac{(F^2)'}{2}. \tag{4.102}$$

An integration yields

$$\eta F' = 2F\left(1 - \frac{F}{4}\right). \tag{4.103}$$

By separation of variables and a second integration we finally obtain (Schlichting 1979)

$$F = \frac{\eta^2}{1 + \eta^2/4}, \tag{4.104}$$

where the integration constants occurring in the integration of Eqs. (4.103) and (4.104) are again determined with the help of Eq. (4.100) to be zero. With Eqs. (4.104) and (4.97) the integral momentum balance given by Eq. (4.81) can now be evaluated and provides the constant c to determine the equation

$$K = \frac{16\pi c K}{3} \quad \text{and} \quad c = \frac{3}{16\pi}. \tag{4.105}$$

Thus, finally, the velocity components u and v are given by

$$u = \frac{3}{16\pi} \frac{K}{\nu x} \frac{2}{(1 + \eta^2/4)^2} \tag{4.106}$$

and

$$v = \sqrt{\frac{3K}{16\pi}} \frac{1}{x} \frac{\eta - \eta^3/4}{(1 + \eta^2/4)^2} \tag{4.107}$$

and for the flow rate we obtain

$$\dot{V} = 2\pi \int_0^\infty u r dr = 8\pi \nu x. \tag{4.108}$$

We finally want to calculate the temperature distribution in the jet. We denote the temperature in the jet center with $\vartheta_c = \vartheta(\mathrm{x}, \eta = 0)$ and assume that a similarity solution for the dimensionless temperature difference exists in the form

$$\frac{\vartheta - \vartheta_\infty}{\vartheta_c - \vartheta_\infty} = G(\eta). \tag{4.109}$$

With Eqs. (4.97) and (4.109) we can rewrite the integral energy balance given by Eq. (4.82) as

$$H = 2\pi(\vartheta_c - \vartheta_\infty)\nu x \int_0^\infty F' G d\eta. \tag{4.110}$$

Using the abbreviation

$$\int_0^\infty F' G d\eta = b \tag{4.111}$$

we obtained the axial dependence of the temperature in the jet center

$$\vartheta_c - \vartheta_\infty = \frac{H}{2\pi b \nu \mathrm{x}} \tag{4.112}$$

and therefore

$$\vartheta - \vartheta_\infty = \frac{H}{2\pi b \nu \mathrm{x}} G(\eta), \tag{4.113}$$

where b is given by Eq. (4.111) to a certain value. Substituting Eqs. (4.113) and (4.97) into Eq. (4.85) yields the differential equation

$$-(FG)' = \frac{1}{\mathrm{Pr}}(\eta G')'$$

whose integration with the boundary condition

$$\eta = 0 : \quad G = 1 \tag{4.114}$$

yields the solution

$$G = \left(1 + \eta^2/4\right)^{-2\,\mathrm{Pr}}. \tag{4.115}$$

The constant b can be finally obtained by evaluating the integral in Eq. (4.111) to

$$\int_0^\infty F' G d\eta = \frac{2}{2\,\mathrm{Pr} + 1} = b. \tag{4.116}$$

Thus, the temperature field in the jet is given by

$$\vartheta - \vartheta_\infty = \frac{H}{2\pi \nu x} \frac{2}{2\,\mathrm{Pr} + 1} \frac{1}{(1 + \eta^2/4)^{2\,\mathrm{Pr}}}. \tag{4.117}$$

From Eq. (4.93), as an estimate for the Reynolds number we obtain

$$\frac{UL}{\nu} \approx \frac{K}{\nu^2}, \tag{4.118}$$

showing that the Reynolds number is of the order of 1 only at very small momentum K. That may serve as an indication that in practice, the jet is laminar only in the vicinity of the orifice, and becomes turbulent for larger distances. For turbulent flow, the flow parameters are independent from the kinematic viscosity ν, so that then

$$u = f(x, r, K) \tag{4.119}$$

and from pure dimension analytical considerations one obtains

$$u = \frac{\sqrt{K}}{x} f\left(\frac{r}{x}\right). \tag{4.120}$$

As a measure of the jet width, we introduce the radius r_0 at which the speed has decreased to half the value of the jet center. It is then

$$r_0 = \text{const.} x, \tag{4.121}$$

and from measurements one obtains (see Schlichting 1979)

$$\text{const.} = 0.088. \tag{4.122}$$

For the flow rate in a corresponding manner we apply

$$\dot{V} = f(K, x) \tag{4.123}$$

and thus obtain

$$\dot{V} = \sqrt{K} x \,\text{const.} \tag{4.124}$$

For a turbulent jet the flow rate is therefore proportional to the square root of the kinematic momentum, while for a laminar jet it increases proportional to the kinematic viscosity, as it was visible in Eq. (4.108). For both laminar and turbulent jet the velocity decreases in the jet center proportional to $1/x$ and the jet width increases in both cases proportional to x. However, the laminar jet at high momentum is slim and the speed is high, while the jet at small momentum is broad with small velocity. For the turbulent jet however, the jet width is independent from the kinematic momentum.

4.6 Laminar Buoyant Jet

As a final example for the class of similarity solutions, we consider the case of laminar buoyant jets (as shown in Fig. 4.6) (see Schlichting 1979). Buoyant jets arise over locally hot surfaces, or when air exits from a chimney with a temperature higher than the ambient temperature. In these cases, the hot fluid flows upwards opposed to gravity as a result of temperature and density differences. In contrast to the momentum driven jet regarded in the previous section here the difference in temperature and the enthalpy of the jet are the driving forces, thus

$$H = 2\pi \int_0^\infty ur(\vartheta - \vartheta_\infty)\mathrm{d}r = \text{const.} \tag{4.125}$$

Setting the x-coordinate vertically upwards against gravity, we obtain the continuity, momentum and heat balance in the form

$$\frac{\partial}{\partial x}(ur) + \frac{\partial}{\partial r}(vr) = 0, \tag{4.126}$$

$$u\frac{\partial u}{\partial x} + v\frac{\partial u}{\partial r} = \frac{\nu}{r}\frac{\partial}{\partial r}\left(r\frac{\partial u}{\partial r}\right) + g\beta\Delta\vartheta \tag{4.127}$$

and

$$u\frac{\partial \vartheta}{\partial x} + v\frac{\partial \vartheta}{\partial r} = \frac{\nu}{\mathrm{Pr}}\frac{1}{r}\frac{\partial}{\partial r}\left(r\frac{\partial \vartheta}{\partial r}\right). \tag{4.128}$$

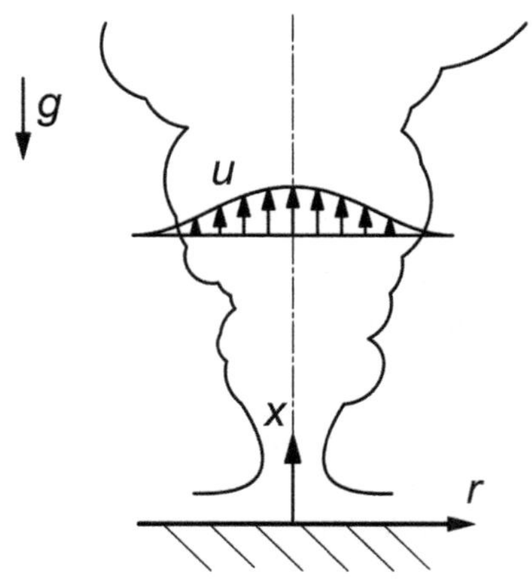

Fig. 4.6 Buoyant jet

The corresponding boundary conditions are given by

$$r = 0: \quad v = 0, \quad \frac{\partial u}{\partial r} = 0, \quad \frac{\partial \vartheta}{\partial r} = 0, \tag{4.129}$$

and

$$r \to \infty: \quad u \to 0, \quad \vartheta \to \vartheta_\infty. \tag{4.130}$$

We again denote the temperature in the jet center by $\vartheta_c = \vartheta(x, r = 0)$, the temperature difference between jet center and the ambient by $\Delta\vartheta = \vartheta_c - \vartheta_\infty$, the characteristic axial length L and the characteristic jet width with R. The magnitudes of the inertial and friction terms then are

$$u \frac{\partial u}{\partial x} \approx \frac{U^2}{L} \tag{4.131}$$

respectively

$$\frac{\nu}{r} \frac{\partial}{\partial r} \left(r \frac{\partial u}{\partial r} \right) \approx \frac{\nu U}{R^2}, \tag{4.132}$$

and from the equality of the magnitude of both terms, we obtain

$$\frac{U^2}{L} \approx \frac{\nu U}{R^2}. \tag{4.133}$$

The integral energy balance given by Eq. (4.125) yields the estimate

$$U R^2 \Delta\vartheta \approx H, \tag{4.134}$$

from which, together with the equality of the magnitude of the friction and the buoyancy forces in Eq. (4.127) it follows that

$$\frac{\nu U}{R^2} \approx g\beta\Delta\vartheta \tag{4.135}$$

and the magnitude of the velocity is obtained as

$$U \approx \sqrt{\frac{g\beta H}{\nu}}. \tag{4.136}$$

Since the inertial and buoyancy forces are of the same order of magnitude we obtain

$$\frac{U^2}{L} \approx g\beta\Delta\vartheta. \tag{4.137}$$

The temperature difference between the jet center and the surrounding ambient results as a suitable measure to

$$\Delta\vartheta \approx \frac{H}{\nu L}, \tag{4.138}$$

and finally from Eq. (4.134)

$$R \approx \left(\frac{\nu^3 L^2}{g\beta H}\right)^{1/4}. \tag{4.139}$$

Again a measure for the length L is missing, so that we have to choose $L = x$. Because of $\psi \approx UR^2$ with Eqs. (4.136) and (4.139) we obtain $\psi \approx \nu L$, and we find

$$\psi = \nu x F(\eta) \quad \text{and} \quad \frac{\vartheta - \vartheta_\infty}{\vartheta_c - \vartheta_\infty} = G \tag{4.140}$$

with

$$\eta = \frac{r}{\sqrt{x}}\left(\frac{g\beta H}{\nu^3}\right)^{1/4}. \tag{4.141}$$

From Eqs. (4.140), (4.141) and (4.88) the velocity components follow

$$u = \sqrt{\frac{g\beta H}{\nu}}\frac{F'}{\eta} \quad \text{and} \quad v = \frac{\nu}{\sqrt{x}}\left(\frac{g\beta H}{\nu^3}\right)^{1/4}\left(\frac{F'}{2} - \frac{F}{\eta}\right). \tag{4.142}$$

Substituting Eqs. (4.140) and (4.142) into the integral heat balance given by Eq. (4.125), one obtains

$$H = 2\pi \int_0^\infty u\Delta\vartheta r \mathrm{d}r = 2\pi\sqrt{\frac{g\beta H}{\nu}}(\vartheta_c - \vartheta_\infty)\int_0^\infty F' G \mathrm{d}\eta, \tag{4.143}$$

and with the abbreviation

$$\int_0^\infty F' G \mathrm{d}\eta = b \tag{4.144}$$

the axial temperature decay in the jet center is given by

$$(\vartheta_c - \vartheta_\infty) = \frac{1}{x}\frac{H}{2\pi b \nu}. \tag{4.145}$$

The ordinary differential equations for the velocity and temperature distributions take the form (see Schlichting 1979)

$$-\left(\frac{FF'}{\eta}\right)' = \left(F'' - \frac{F'}{\eta}\right)' + \frac{\eta G}{2\pi b} \tag{4.146}$$

and

$$-FG = \frac{1}{\Pr}\eta G'. \tag{4.147}$$

The flow rate thus is again obtained from the integral

$$\dot{V} = 2\pi \int_0^\infty ur\mathrm{d}r = 2\pi\nu x \int_0^\infty F'\eta \mathrm{d}\eta. \tag{4.148}$$

From Eq. (4.136) the Reynolds number can be estimated to be

$$\frac{UL}{\nu} \approx \sqrt{\frac{g\beta H}{\nu^3}}x, \tag{4.149}$$

showing that the Reynolds number becomes larger with increasing distance. We therefore expect that also the buoyant jet is laminar only in the immediate vicinity of the origin, and becomes turbulent at larger distances from the origin.

For turbulent flow the variables are again independent of the kinematic viscosity ν and also independent of α, so that then

$$u = f(x, r, g\beta, H), \tag{4.150}$$

and from dimension analytical considerations one obtains

$$u = \left(\frac{g\beta H}{x}\right)^{1/3} f\left(\frac{r}{x}\right). \tag{4.151}$$

The width r_0 of the jet at which the speed has dropped to half the value of the jet center is given by

$$r_0 = \text{const.}x, \tag{4.152}$$

where now by measurements one finds

$$\text{const.} = 0.11. \tag{4.153}$$

For the temperature profile in the jet center the analogous relationship

$$\vartheta_c - \vartheta_\infty = f(x, r, g\beta, H) \tag{4.154}$$

is obtained, yielding

$$\vartheta_c - \vartheta_\infty = \left(\frac{H^2}{g\beta x^5}\right)^{1/3} f\left(\frac{r}{x}\right). \tag{4.155}$$

For the flow rate in a corresponding manner it follows

$$\dot{V} = f(x, g\beta, H). \tag{4.156}$$

and thus

$$\dot{V} = \left(g\beta H x^5\right)^{1/3} \text{const.} \tag{4.157}$$

For the buoyant jet the momentum is zero at the source and increases with x. From

$$K = f(x, g\beta, H) \tag{4.158}$$

one obtains

$$K = \left(g\beta H x^2\right)^{2/3} \text{const.} \tag{4.159}$$

For buoyant jets the exiting momentum is often no longer negligible, and the jet behavior depends on both H and K. With H and K, a typical length

$$\lambda = \frac{K^{3/4}}{(g\beta H)^2} \tag{4.160}$$

can be formed. With this length x and r can be rendered dimensionless, and thus the similarity is disturbed in the velocity and temperature distributions. Since the velocity of the free jet decreases with x^{-1} and that of the buoyant jet with $x^{-1/3}$, the jet initiates as a slim jet ($r_0/x = 0.088$) and then changes over to a broader buoyant jet ($r_0/x = 0.11$), Spurk (1992).

Appendix A
Selected Problems

Appendix A provides a set of selected problems. The solutions to these problems are summarized in Appendix B. Appendix A starts with some introductory problems, followed by problems which need additional considerations for obtaining a solution. Furthermore some problems are devoted to the choice of the base system and to model theory and also on similarity solutions. The problems are thought to provide a good basis for exercises in order to very well understand the beauty of dimensional analysis.

A.1 Problems Which Lead to Optimal Results

A1.1 Speed of Sound in Gases

The state of a gas is characterized by the pressure p and density ρ. Determine the equation for the speed of sound a, using dimensional analysis.

A1.2 Speed of Sound in Elastic Bodies

An elastic body shall have the elasticity modulus E and density ρ. Determine the speed of sound a in dimensionless form.

A1.3 Natural Frequency of a Taut String

The natural frequency n of a string (e.g. the string of a guitar or a violin) depends on the force F with which the string is strained, its density q (in this case: mass per unit length) and the length of the string L. Determine the natural frequency n of the string using dimensional analysis.

V. Simon et al., *Dimensional Analysis for Engineers*,
Mathematical Engineering, DOI 10.1007/978-3-319-52028-5

A1.4 Viscosity of Gases

The macroscopically observed property of viscosity of gases originates in the thermal motion of the molecules. The molecular motion is affected by the molecular weight m, the average molecular velocity v and the intermolecular force F written in the form

$$F = K\,r^{-n}, \quad K, n = \text{const.},$$

whereas r denotes the distance between the molecules. According to kinetic theory of gases, the temperature is proportional to the average kinetic energy of the molecule, i.e.

$$\left\langle \frac{mv^2}{2} \right\rangle = k_B \vartheta,$$

with the Boltzmann constant k_B. Determine the dependency between the gas viscosity and its temperature.

A.2 Problems That Require Additional Considerations

A2.1 Oscillations of a Pendulum

At the end of a pendulum of length L a mass m is attached (see Fig. A.1). Determine with the help of dimensional analysis, the formula for the period of the oscillations τ of the pendulum, for very small initial displacements φ_0.

A2.2 Velocity of a Bubble Rising Upward

An upright standing tube of diameter d is filled with a liquid of density ρ_w. In the liquid a bubble with the density ρ_L rises upwards with the speed U. It can be

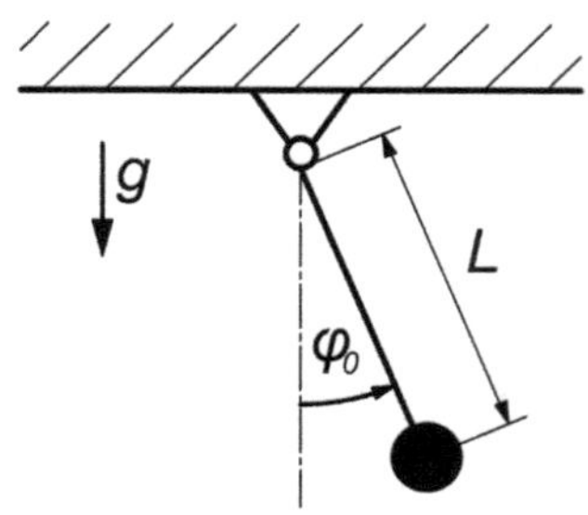

Fig. A.1 Mathematical pendulum

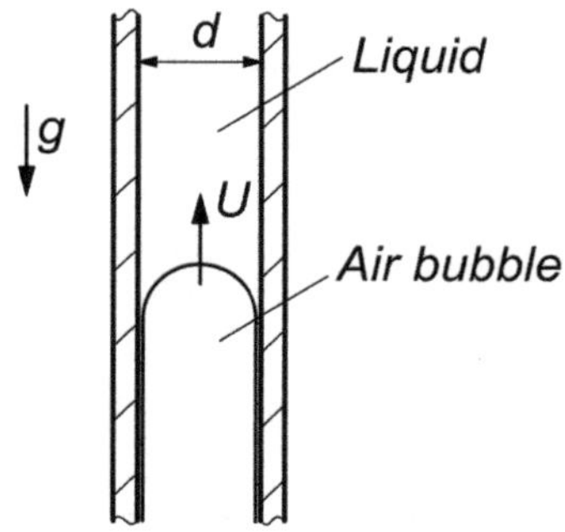

Fig. A.2 Bubble in a tube filled with liquid

assumed that the air bubble almost entirely fills the circular cross section of the pipe and is very long (Fig. A.2). Determine the rising velocity of the air bubble U by dimensional analysis. At first specify the dimension matrix, and determine its rank and the number of dimensionless products. Deduce the dimensionless products, and discuss the special case $\rho_L/\rho_w \ll 1$.

A2.3 Capillary Height in a Tube

At the interface between two liquids (such as water and air) always surface tension occurs. In still liquid, the pressure difference between two fluids at the separating interface is proportional to the surface tension and the mean curvature. In general, the separating surface between two fluids inflects at the edges, when two liquids—for example water and air—touch a solid wall. The angle α enclosed between interface and the wall is a function of the considered liquids and the wall material (see Fig. A.3). The interface bends and a pressure difference develops between the liquids. In capillary tubes with small diameters, the pressure difference causes the liquid in the tube to rise upwards against gravity or to be pressed downwards. The height difference h in the tube of diameter d thus is a function of surface tension σ and the liquid density ρ, the gravitational acceleration g and the angle α. Determine, the height h using dimensional analysis.

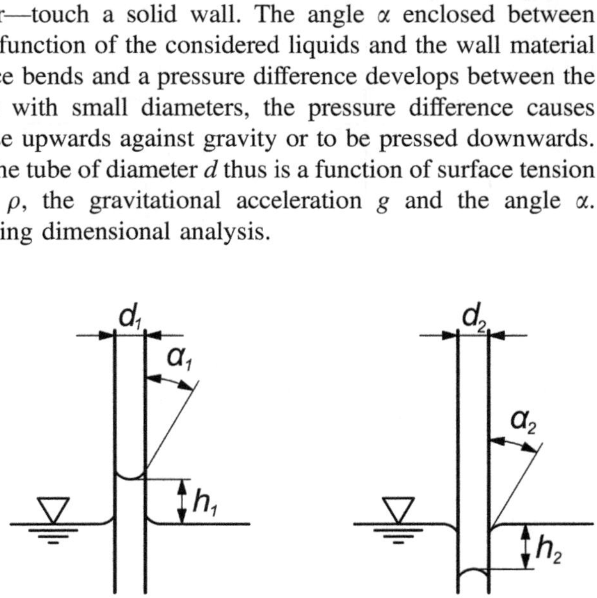

Fig. A.3 Capillary tube for water (*left*) and mercury (*right*)

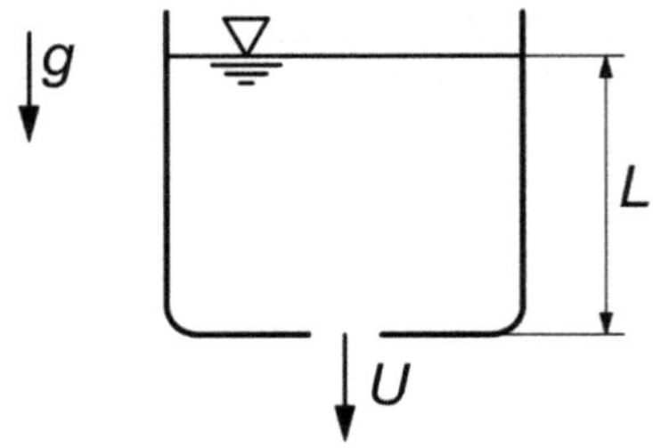

Fig. A.4 Liquid outflow from a large container

A2.4 Liquid Outflow from a Large Container

A large container is filled with liquid of density ρ and viscosity η. This is schematically shown in Fig. A.4. The liquid is released at the bottom of the container through an opening of diameter d at speed u. Determine the outflow velocity of the liquid.

A2.5 Volume Flow Measurement at Open Water Courses

To measure the volume flux at open water courses a vertical wall is used that has an opening, through which the water passes (see Fig. A.5). For a rectangular opening of width b the flow rate depends not only on the width but also on the water level h to the bottom edge of the opening and the forcing of gravity. Determine the volume flow rate.

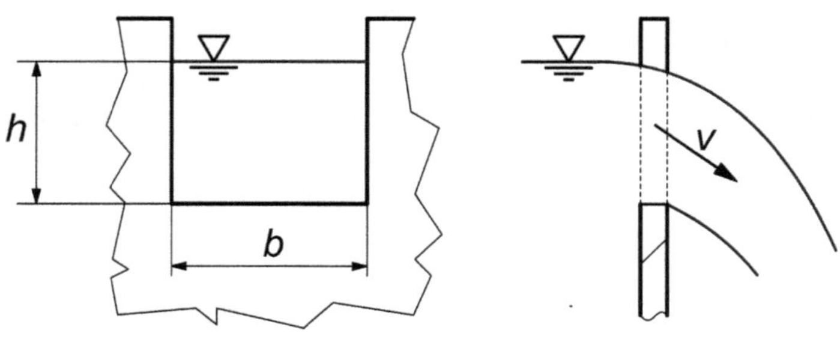

Fig. A.5 Volume flow measurement

A.3 Problems Related with the Choice of the Base System

A3.1 Pressure Drop in a Pipe

An incompressible fluid with density ρ and viscosity ηflows through a straight circular pipe of diameter d. The pressure gradient is given by $K = -\Delta p/L$. Determine the mass flow $\dot{m}$ using the $[LMFT]$ system and discuss the result.

A3.2 Rotating Cylinder

An infinitely long cylinder of radius R is located in a liquid with viscosity η and density ρ and rotates with the angular velocity Ω around its axis. Determine the moment M that acts on the cylinder by dimensional analysis in the $[LMFT]$system. Determine the dimension matrix, and determine its rank and the number of dimensionless products. Specify the dimensionless products, and discuss both the general case as well as the special case occurring when the dimensional constant C is not considered in the $[LMFT]$ system.

A.4 Problems to Exercise the Formal Determination of Dimensionless Products

A4.1 Resistance of a Body Under Water

A sphere of diameter d moves with the velocity U through an incompressible fluid with density ρ and viscosity η. This is shown in Fig. A.6. Determine the resistance of the sphere when it moves at a certain distance h below the free surface of the liquid including the surface tension σ. The influence of the atmosphere can be neglected.

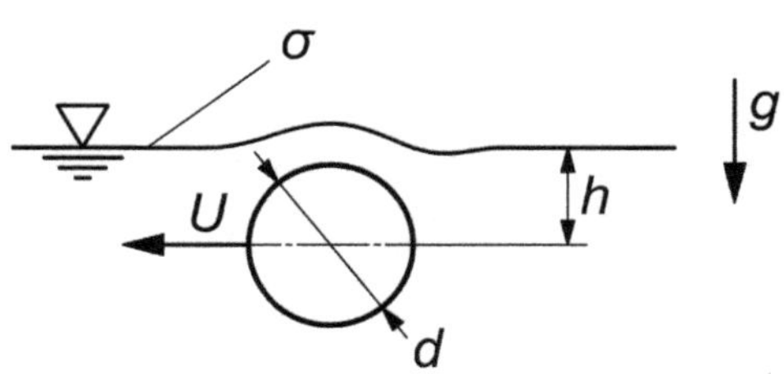

Fig. A.6 Resistance of a body under water

A.5 Problems Related to Model Theory

A5.1 Buckling Load of a Steel Column

In order to determine the buckling load of a steel column ($E = 21.1 \times 10^{10}\ \mathrm{N/mm^2}$), a load test is conducted on a ten times smaller, geometrically similar model from aluminum ($E = 0.7 \times 10^{10}\ \mathrm{N/mm^2}$). Figure A.7 shows a schematic of the experiment. The experiments reveal a buckling load $P_{kr} = 2.5$ kN of the model. What is the critical load of the original element?

A5.2 Deformation and Stress of a Cantilever Beam

A cantilever that is anchored at one end and carries a load P at the other end is considered (see Fig. A.8). A small scale model is investigated, to draw conclusions on the maximum deflection and the stresses in the beam of the full size version. Specify the required scale factors, if the scale factor is $M_L = 0.1$ for the beam length.

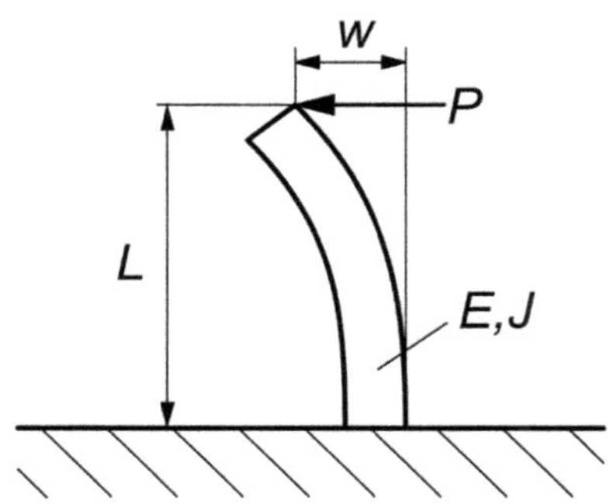

Fig. A.7 Buckling load of a steel column

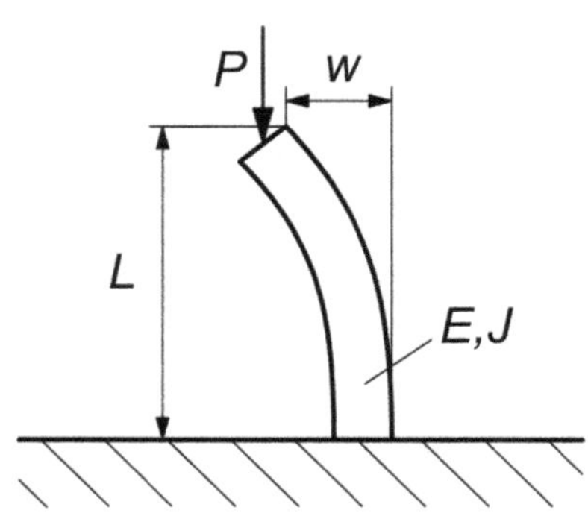

Fig. A.8 Deformation of a cantilever beam

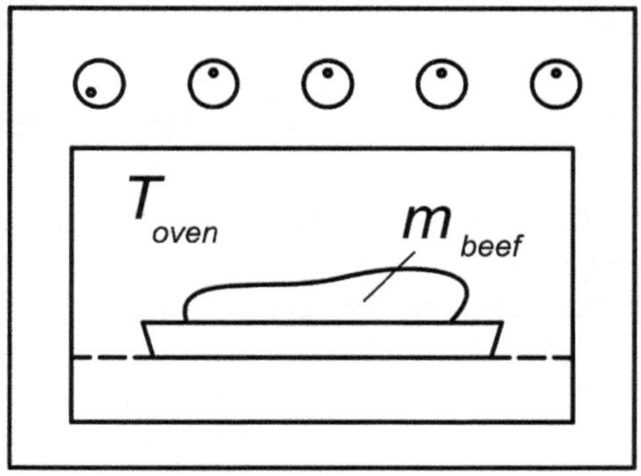

Fig. A.9 Piece of roast beef in an oven

A5.3 Cooking of Roast Beef

Following a cookbook recipe (for four people), a 1 kg heavy piece of roast beef must fry in a preheated 175 °C oven for 1 h (see Fig. A.9). Because eight guests are expected, the chef doubles the piece of meat to 2 kg. How long must the piece roast with geometric similarity at the same oven temperature?

A.6 Problems which lead to Similarity Solutions of Partial Differential Equations

A6.1 Vortex Decay

In an incompressible fluid of density ρ and viscosity η an infinitely long straight vortex filament with constant circulation Γ. The velocity field induced by the vortex has only one component v in the circumferential direction, while the radial and axial velocity components are zero. At time $t = 0$, the velocity field is given by

$$v(r, t = 0) = \frac{\Gamma}{2\pi r}. \tag{A.1}$$

Without further supplying energy to the vortex, as a results of the fluid viscosity, it decays. Besides the initial condition (A.1) the velocity field must satisfy the boundary conditions

$$v(r = 0, t) = 0 \tag{A.2}$$

and

$$v(r \to \infty, t) = \frac{\Gamma}{2\,\pi\,r}. \tag{A.3}$$

This flow identically satisfies the continuity equation and the axial and radial components of the equation of motion. The circumferential component of the momentum equation simplifies to

$$\frac{\partial v}{\partial t} = \frac{\eta}{\rho}\left(\frac{\partial^2 v}{\partial r^2} + \frac{1}{r}\frac{\partial v}{\partial r} - \frac{v}{r^2}\right). \tag{A.4}$$

(a) Determine the dimension matrix for the velocity distribution $v = v(r, t, \Gamma, \eta, \rho)$ and determine the dimensionless products.
(b) Show that the partial differential equation (A.4) can be converted into an ordinary differential equation using the dimensionless products obtained in (a), and determine the solution of this differential equation that satisfies initial and boundary conditions according to Eqs. (A.1), (A.2) and (A.3).

Appendix B
Solutions to the Problems

Appendix B provides the solution of all problems given in Appendix A. The solution is presented in a way to show all the major steps in the solution process of the problem.

Problem A1.1 The speed of sound a in gas depends on the local pressure p, and the local density ρ. It has therefore a relationship of the form

$$a = f(p, \rho). \tag{B.1}$$

The variables describing the problem are summarized in the following table.

Variable	Symbol	Dimension formula
Speed of sound	a	$\mathrm{L\,T^{-1}}$
Pressure	p	$\mathrm{L^{-1}\,M\,T^{-2}}$
Density	ρ	$\mathrm{L^{-3}\,M}$

The product p/ρ has the dimension

$$\left[\frac{p}{\rho}\right] = \mathrm{L^2\,T^{-2}} \quad \text{and the unit} \quad \left\{\frac{p}{\rho}\right\} = \mathrm{m^2\,s^{-2}}, \tag{B.2}$$

that is, with $\left[\sqrt{p/\rho}\right] = LT^{-1}$ the speed of sound can be made dimensionless, and we obtain

$$\prod = a\sqrt{\frac{\rho}{p}} \quad \text{respectively} \quad a\sqrt{\frac{\rho}{p}} = \text{const.} \tag{B.3}$$

V. Simon et al., *Dimensional Analysis for Engineers*,
Mathematical Engineering, DOI 10.1007/978-3-319-52028-5

The speed of sound is then

$$a = \text{const.}\sqrt{\frac{p}{\rho}}. \tag{B.4}$$

This relation holds strictly only for calorically perfect gases, for which const. $= \sqrt{\gamma}$, where γ is the isentropic exponent of the gas.

Problem A1.2 The speed a at which sound propagates in an elastic body, is dependent on the elasticity modulus E and the density ρ of the body. It is therefore

$$a = \mathrm{f}(E, \rho). \tag{B.5}$$

Thus, one can formulate the following table.

Variable	Symbol	Dimension formula
Speed of sound	a	$\mathrm{L\,T^{-1}}$
Modulus of elasticity	E	$\mathrm{L^{-1}\,M\,T^{-2}}$
Density	ρ	$\mathrm{L^{-3}\,M}$

The product E/ρ has the dimension of a velocity again

$$\left[\frac{E}{\rho}\right] = \mathrm{L^2T^{-2}} \quad \text{with the unit} \quad \left\{\frac{E}{\rho}\right\} = \mathrm{m^2\,s^{-2}}, \tag{B.6}$$

so that the speed of sound can be non-dimensionalized with $\sqrt{E/\rho}$ yielding

$$a = \text{const.}\sqrt{\frac{E}{\rho}}. \tag{B.7}$$

The theory provides const. $= 1$. If in addition to the sound propagation also a displacement velocity occurs as a result of an external force, the result is the so called Cauchy number

$$\mathrm{Ca} = \left(\frac{v}{a}\right)^2. \tag{B.8}$$

In the theory of elasticity the Cauchy number plays an analogous role as the Mach number in fluid mechanics.

Problem A1.3 The natural frequency n of the string is dependent on the force F with which the string is biased, the density q (in this case, mass per unit length) and the length L of the string. The unknown function thus has the form

$$n = f(F, q, L), \tag{B.9}$$

and the physical quantities are summarized in the following table of dimension formulas.

Variable	Symbol	Dimension formula
Natural frequency	n	T^{-1}
Force	F	$\mathrm{L\,M\,T}^{-2}$
Density	q	$\mathrm{L}^{-1}\,\mathrm{M}$
Length	L	L

The only dimensionless product is thus

$$nL\sqrt{\frac{q}{F}} = \text{const.}, \tag{B.10}$$

and we obtain the equation for the desired natural frequency as

$$n = \text{const.}\,\frac{1}{L}\sqrt{\frac{F}{q}}, \tag{B.11}$$

from which one deduces that the frequency increases as the force increases and decreases with increasing density. The solution of the corresponding physical equation yields the value of the constant to const. $= 1/2$.

Problem A1.4 The macroscopically observed property of the viscosity of gases is originated in the thermal motion of the molecules that causes the smearing of velocity gradients. As a result, a momentum exchange takes place so that from a continuum theoretical point of view the viscosity η becomes noticeable. Except for the molecular weight m and the mean molecular velocity v, the molecular motion is also influenced by the intermolecular forces. Molecules of a gas repel each other with a force F that decreases with increasing distance r between the molecules that can be written in the form

$$F = K\,r^{-n}, \quad K, n = \text{const.}, \tag{B.12}$$

where the constant K has the dimension $\mathrm{L}^{1+n}\,\mathrm{M}\,\mathrm{T}^{-2}$ and alike the exponent n is a characteristic property of the considered molecules. We now ask for the viscosity of gases as a function of the molecular mass m, the speed v and the force F and expect a correlation of the form

$$\eta = f(m, v, F). \tag{B.13}$$

Within dimensional analysis, the force F is described by the dimensionless constant K, whose dimension contains the exponent n and thus all necessary information on the forcing between the molecules. Equation (B.13) thus takes the form

$$\eta = f(m, v, K). \tag{B.14}$$

Furthermore, the dependency of viscosity on temperature, which is proportional to the average kinetic energy of a molecule according to the kinetic theory of gases, is given by

$$\left\langle \frac{m\,v^2}{2} \right\rangle = k_B\,\vartheta. \tag{B.15}$$

Here, k_B denotes the Boltzmann constant. Since the mass of the molecule already appears in the argument of the function, we will replace the speed by the temperature and the Boltzmann constant, and write Eq. (B.14) in the form

$$\eta = f(m, \vartheta, K, k_B). \tag{B.16}$$

The dimension matrix is then

	η	m	ϑ	K	k_B
L	−1	0	0	$n+1$	2
M	1	1	0	1	1
T	−1	0	0	−2	−2
θ	0	0	1	0	−1

and may be transformed into the form

	η	m	$k_B\,\vartheta$	K	k_B
L	−1	0	2	$n+1$	2
M	1	1	1	1	1
T	−1	0	−2	−2	−2
θ	0	0	0	0	−1

From which one deduces that k_B may no longer appear in the problem, since no further variable contains the dimension of temperature. Further rearrangement results to

	η/m	m	$k_B\,\vartheta/K$	K
L	−1	0	$1-n$	$n+1$
M	0	1	0	1
T	−1	0	0	−2

and thus

	η/m	m	$k_B\,\vartheta/K$	K/m
L	−1	0	$1-n$	$n+1$
M	0	1	0	0
T	−1	0	0	−2

so that the mass cannot enter the problem and thus yielding

	$\eta/m\,(m/K)^{1/2}$	$k_B\,\vartheta/K$	K/m
L	$-(n+3)/2$	$1-n$	$n+1$
M	0	0	0
T	0	0	−2
θ	0	0	0

The only dimensionless product is therefore

$$\frac{\eta}{m}\left(\frac{m}{K}\right)^{1/2}\left(\frac{k_B\,\vartheta}{K}\right)^{(n+3)/(2-2n)} = \text{const.} \tag{B.17}$$

that solved for the viscosity becomes

$$\eta = \text{const.}(K\,m)^{1/2}(k_B\,\vartheta/K)^{(n+3)/(2n-2)}. \tag{B.18}$$

Thus, for $n = 5$ the viscosity is directly proportional to the temperature, while in the limiting case $n \to \infty$ the viscosity increases as the square root of the temperature.

Problem A2.1 The initial deflection of the pendulum is $\varphi(t=0)=\varphi_0$. The requested period τ is a function of

$$\tau = f(L, m, g, \varphi_0). \tag{B.19}$$

The dimension matrix then takes the form

	τ	L	m	g	φ_0
L	0	1	0	1	0
M	0	0	1	0	0
T	1	0	0	−2	0

Since only m contains the dimension of mass, no combination of parameters can be found to render the quantity dimensionless. Therefore, the mass m does not enter the problem, and the corresponding row can be deleted from the dimension matrix. Besides the already non-dimensional initial displacement φ_0, the remaining variables the dimensionless product

$$\Pi = \tau\sqrt{\frac{g}{L}} \tag{B.20}$$

yielding the relation

$$f\left(\tau\sqrt{\frac{g}{L}}, \varphi_0\right) = 0 \quad \text{respectively} \quad \tau\sqrt{\frac{g}{L}} = f(\varphi_0). \tag{B.21}$$

So far only dimension analytical approaches are applied. Further progress requires additional considerations. The function $f(\varphi_0)$ must be an even function, i.e.

$$f(\varphi_0) = f(-\varphi_0). \tag{B.22}$$

A Taylor expansion around $\varphi_0 = 0$ therefore must have the form

$$f(\varphi_0) = a_0 + a_2\,\varphi_0^2 + a_4\,\varphi_0^4 + \cdots \tag{B.23}$$

with constant coefficients a_i. Neglecting terms which are quadratic in φ_0, we obtain

$$f(\varphi_0) \doteq a_0 \tag{B.24}$$

and thus

$$\tau\sqrt{\frac{g}{L}} = a_0 \quad \text{for} \quad \varphi_0 \to 0, \tag{B.25}$$

respectively

$$\tau = a_0\sqrt{\frac{L}{g}} \quad \text{for} \quad \varphi_0 \to 0. \tag{B.26}$$

The equation of motion for small amplitudes around the idle state

$$\ddot{\varphi} + \frac{g}{L}\varphi = 0 \tag{B.27}$$

yields the natural frequency

$$\omega = \sqrt{\frac{g}{L}} \tag{B.28}$$

and the period

$$\tau = \frac{2\,\pi}{\omega} = 2\,\pi\sqrt{\frac{L}{g}} \tag{B.29}$$

so that the constant a_0 is determined to $2\,\pi$.

Problem A2.2 We consider an upright standing cylinder of diameter d filled with a liquid of density ρ_w. Under certain circumstances an air bubble with the lower density ρ_L is formed and rises upwards at the speed U. We assume that the air bubble entirely fills the circular cross section of the pipe and is very long. The speed of the rising bubble will depend on the pipe diameter d, the densities ρ_w and ρ_L as well as the acceleration of gravity g so that

$$U = f(d, g, \rho_w, \rho_L). \tag{B.30}$$

The dimension matrix then reads

	U	d	g	ρ_w	ρ_L
L	1	1	1	−3	−3
M	0	0	0	1	1
T	−1	0	−2	0	0

yielding the nondimensional products

$$\prod_1 = \frac{U}{\sqrt{gd}} \quad \text{and} \quad \prod_2 = \frac{\rho_L}{\rho_w}. \tag{B.31}$$

so that the relationship may be written as

$$f\left(\prod_1, \prod_2\right) = 0 \quad \text{respectively} \quad \prod_1 = f\left(\prod_2\right) \tag{B.32}$$

or

$$\frac{U}{\sqrt{gd}} = f\left(\frac{\rho_L}{\rho_w}\right). \tag{B.33}$$

In the limiting case that $\rho_L/\rho_w << 1$ one may simplify the relation to

$$f\left(\frac{\rho_L}{\rho_w}\right) = f(0) = \text{const.}, \tag{B.34}$$

where the constant value follows from measurements to const. = 0.35.

Problem A2.3 The height h to which the liquid rises in a capillary tube of diameter d, is a function of the surface tension σ and the density, the gravitational acceleration g and the angle α, according to

$$h = f(d, \sigma, g, \rho, \alpha). \tag{B.35}$$

The dimension matrix thus reads

	h	d	σ	g	ρ	α
L	1	1	0	1	−3	0
M	0	0	1	0	1	0
T	0	0	−2	−2	0	0

and can be transformed to

	h/d	d	σ/ρ	g	ρ	α
L	0	1	3	1	−3	0
M	0	0	0	0	1	0
T	0	0	−2	−2	0	0

and subsequently to

	h/d	d	$\sigma/(\rho g)$	g	α
L	0	1	2	1	0
M	0	0	0	0	0
T	0	0	0	−2	0

where in the first step the dimension of mass is eliminated by combining the gravitational acceleration and the density. The product

$$a = \sqrt{\frac{\sigma}{\rho g}} \tag{B.36}$$

possesses the dimensions of a length and is often referred to as Laplace length. Besides gravitational acceleration it only depends on material properties. For water at 20 °C it takes the value 0.39 m. The dimensionless products are

$$\Pi_1 = \frac{h}{d}, \quad \Pi_2 = \frac{\sigma}{\rho g d^2}, \quad \Pi_3 = \alpha, \tag{B.37}$$

yielding the relation

$$\frac{h}{d} = f\left(\frac{\sigma}{\rho g d^2}, \alpha\right) \tag{B.38}$$

or

$$\frac{h}{d} = f\left(\frac{a}{d}, \alpha\right). \tag{B.39}$$

Further improvement requires additional physical insight. From observation, we know that the height increases as the diameter of the tube becomes smaller. Assuming that $h \propto 1/d$ or that the height occurs only in the combination hd, besides α the dimensionless product

$$\Pi_1^* = \Pi_1 \Pi_2^{-1} = \frac{\rho g h d}{\sigma} \tag{B.40}$$

is obtained and the functional relation can be transformed to

$$\frac{\rho g h d}{\sigma} = f(\alpha). \tag{B.41}$$

By analytical means one finds $f(\alpha) = 4\cos(\alpha)$.

Problem A2.4 A large container contains a liquid of density ρ and viscosity η. The liquid is released at the bottom of the container through an opening of diameter d at speed u. It is sought for the liquid outflow velocity in the form

$$u = f(h, d, g, \eta, \rho) \tag{B.42}$$

The dimension matrix reads

	u	h	d	g	η	ρ
L	1	1	1	1	−1	−3
M	0	0	0	0	1	1
T	−1	0	0	−2	−1	0

and is reshaped to

	u	h	d/h	gh	η	ρ/η
L	1	1	0	2	−1	−2
M	0	0	0	0	1	0
T	−1	0	0	−2	−1	1

At this point the viscosity must vanish, since no other variable containing the dimension of mass is left. After further transformation, one obtains

	$u/\sqrt{gh}$	h	d/h	gh	$h\rho/\eta$
L	0	1	0	2	−1
M	0	0	0	0	0
T	0	0	0	−2	1

and finally the dimensionless products result to

$$\Pi_1 = \frac{u}{\sqrt{gd}}, \quad \Pi_2 = \frac{\rho h\sqrt{gh}}{\eta} \quad \text{and} \quad \Pi_3 = \frac{d}{h}. \tag{B.43}$$

The relation between the products is then

$$f(\Pi_1, \Pi_2, \Pi_3) = 0 \quad \text{respectively} \quad \Pi_1 = f(\Pi_2, \Pi_3) \tag{B.44}$$

or

$$\frac{u}{\sqrt{gh}} = f\left(\frac{\sqrt{gh}\rho h}{\eta}, \frac{d}{h}\right). \tag{B.45}$$

Additional physical considerations allow further simplification. The product $\prod_2$ can be identified as the Reynolds number. Since it is expected that for large Reynolds numbers, i.e. in the limit of negligible friction, a non-trivial result is obtained, the product $\prod_2$ is retracted, yielding

$$\frac{u}{\sqrt{gh}} = f\left(\frac{d}{h}\right). \tag{B.46}$$

However, even this relationship can be further simplified if one considers that the volume flow $\dot{V}$ is proportional to the outflow area and, therefore, the speed cannot depend on the diameter of the outlet opening. We finally obtain

$$u = \text{const.}\sqrt{gh}, \tag{B.47}$$

whereas theoretical considerations provide the value const. $= \sqrt{2}$. A related problem is given by the sand descending in an hourglass or the outflow of grain from a silo. As long as the grain diameter is comparatively small thus $k \ll d$ and $k \ll h$, the dry sand or the cereal can be well approximated as a continuum. However, other than for liquids it is found, that the pressure at the bottom of the vessel is independent of the filling level h. In fact the sand flows only very close to the opening hole. Close to the opening the sand moves downwards in a funnel-shaped tube, or in a cylindrical tube at larger distances. Outside that tube the sand does not move. As a result of this observation and the independence of the ground pressure from the container height yields a relation of the form

$$u = f(g, d) \tag{B.48}$$

thus

$$u = \text{const.}\sqrt{gd} \tag{B.49}$$

resulting in the relation $\dot{V} \propto d^{5/2}$ for the volume flow, also known as "5/2-law" (see e.g. Wieghardt 1952).

Problem A2.5 The water course is characterized by its width b and the height of the water level h. From the previous example it is apparent, that for calculating the outflow velocity of liquids from large containers, the influence of viscosity can be neglected. But then the density can no longer be part of the solution, as it becomes the only variable of the problem that contains the dimension of mass. Accordingly the sought relation for the volume flow is of the form

$$\dot{V} = f(g, b, h), \tag{B.50}$$

with the corresponding dimension matrix

	$\dot{V}$	g	b	h
L	3	1	1	1
M	0	0	0	0
T	−1	−2	0	0

with

	$\dot{V}/h^2$	gh	b/h	h
L	1	2	0	1
M	0	0	0	0
T	−1	−2	0	0

it follows that besides the length ratio

$$\Pi_1 = \frac{b}{h} \tag{B.51}$$

the combination

$$\Pi_2 = \frac{\dot{V}}{h^2\sqrt{gh}} \tag{B.52}$$

occurs, revealing the functional relation

$$\frac{\dot{V}}{h^2\sqrt{gh}} = f\left(\frac{b}{h}\right). \tag{B.53}$$

An improved statement is obtained by the assumption that the flow is proportional to the width of the opening, so that

$$\frac{\dot{V}}{b} = f(g, h). \tag{B.54}$$

In that case the dimension matrix reads

	$\dot{V}/b$	g	h
L	2	1	1
M	0	0	0
T	−1	−2	0

yielding the only dimensionless product

$$\frac{\dot{V}}{bh\sqrt{gh}} = \text{const.} \tag{B.55}$$

Problem A3.1 An incompressible fluid with the density ρ and viscosity η flows through a straight circular pipe of diameter d. The pressure gradient is given by $K = -\Delta p/L$. The mass flow is then a function of the following form

$$\dot{m} = f(K, \eta, \rho, d). \tag{B.56}$$

The dimension matrix in the $[LMFT]$ system reads

	$\dot{m}$	K	η	ρ	d
L	0	−3	−2	−3	1
M	1	0	0	1	0
F	0	1	1	0	0
T	−1	0	1	0	0

and after transformation it becomes

	$\dot{m}$	K/η	η	ρ	d
L	0	−1	−2	−3	1
M	1	0	0	1	0
F	0	0	1	0	0
T	−1	−1	1	0	0

This matrix can be rewritten, because η can not be part of the solution. This results in

	$\dot{m}/\rho$	$K\,d/\eta$	ρ	d
L	3	0	−3	1
M	0	0	1	0
F	0	0	0	0
T	−1	−1	0	0

Thus, only one dimensionless product is obtained

$$\frac{\dot{m}\eta}{\rho\,K\,d^4} = \text{const.} \tag{B.57}$$

If we solve this equation for the mass flow, we obtain

$$\dot{m} = \text{const.}\frac{\rho K d^4}{\eta}. \tag{B.58}$$

By means of the $[LMFT]$ system corresponding to Newton's law the introduction of a dimensional constant C is required, thus

$$F = Cma. \tag{B.59}$$

Otherwise, Newton's law would not be dimensionally homogeneous. This constant is however not considered in the derivation of Eqs. (B.57) and (B.58), therefore implying that Newton's law is not relevant for the considered problem, i.e. the flow is not accelerated. The solution given by Eq. (B.57) or (B.58) thus corresponds to the laminar Hagen-Poiseuille flow and the constant may be determined from theoretical considerations to be const. $= \pi/128$ (see Schlichting 1979).

To account for the validity of Newton's law in the $[LMFT]$ system, the dependence of the constant C must be included and the function given by Eq. (B.56) is extended to

$$\dot{m} = f(K, \eta, \rho, d, C). \tag{B.60}$$

The dimension matrix becomes

	$\dot{m}$	K	η	ρ	d	C
L	0	−3	−2	−3	1	−1
M	1	0	0	1	0	−1
F	0	1	1	0	0	1
T	−1	0	1	0	0	2

Transformation yields

	$\dot{m}$	K/η	η	ρ	d	C/η
L	0	−1	−2	−3	1	1
M	1	0	0	1	0	−1
F	0	0	1	0	0	0
T	−1	−1	1	0	0	1

This shows that η cannot be part of the solution and one obtains

	$\dot{m}/\rho$	K/η	ρ	d	$C\dot{m}/\eta$
L	3	−1	−3	1	1
M	0	0	1	0	0
F	0	0	0	0	0
T	−1	−1	0	0	0

From the dimension matrix we see that also ρ cannot be part of the solution and thus one obtains

	$\dot{m}\eta/(\rho K d^4)$	K/η	d	$C\dot{m}/(\eta d)$
L	0	−1	1	0
M	0	0	0	0
F	0	0	0	0
T	0	−1	0	0

The two dimensionless products result to

$$\prod_1 = \frac{\dot{m}\eta}{\rho K d^4} \quad \text{and} \quad \prod_2 = \frac{C\dot{m}}{\eta d}. \tag{B.61}$$

Equation (B.60) then becomes

$$\frac{\dot{m}\eta}{\rho K d^4} = f\left(\frac{C\dot{m}}{\eta d}\right). \tag{B.62}$$

Both dimensionless products contain the viscosity. It is however beneficial to choose the products such that the viscosity appears only in a single product, and therefore $\prod_1$ is replaced by

$$\prod_1^* = \prod_1 \prod_2 = \frac{C\dot{m}^2}{\rho K d^5}. \tag{B.63}$$

Further introducing a mean velocity defined by

$$\dot{m} = \rho\,\dot{V} = \rho\,\overline{U}\frac{\pi}{4}d^2 \tag{B.64}$$

results to

$$\prod_2 = C\frac{\rho\overline{U}d}{\eta} = C\,\mathrm{Re} \quad \text{with} \quad \mathrm{Re} = \frac{\rho\overline{U}d}{\eta}. \tag{B.65}$$

To return to the $[LMT]$-system, the constant and its dimension are set to unity, thus $[C] = 1$. From Eq. (B.62) the functional dependence then reads

$$\frac{\dot{m}^2}{\rho\,K\,d^5} = f(\mathrm{Re}) \tag{B.66}$$

and solving for the mass flow reveals

$$\dot{m} = \sqrt{\rho K d^5} f(\mathrm{Re}). \tag{B.67}$$

In contrast to Eq. (B.58), the mass flow is no longer directly proportional to the pressure gradient, but increases proportionally to its square root.

Problem A3.2 The moment M acting on a cylinder of radius R resulting from a rotation with the angular velocity Ω in a liquid with the viscosity η and density ρ follows to

$$M = f(\eta, \rho, \Omega, R). \tag{B.68}$$

In the $[LMFT]$-System the corresponding dimension matrix reads

	M	η	ρ	Ω	R	C
L	1	−2	−3	0	1	−1
M	0	0	1	0	0	−1
F	1	1	0	0	0	1
T	0	1	0	−1	0	2

The dimension matrix is of the rank $r = 4$, so that the number of dimensionless products results to $d = n - r = 2$. After transformation

	M/η	η	ρ	Ω	R	$C\rho/\eta$
L	3	−2	−3	0	1	−2
M	0	0	1	0	0	0
F	0	1	0	0	0	0
T	−1	1	0	−1	0	1

One notices that ρ and η cannot be part of the solution. Thus, we obtain

	$M/(\eta\Omega R^3)$	Ω	R	$C\rho\,\Omega R^2/\eta$
L	0	0	1	0
M	0	0	0	0
F	0	0	0	0
T	0	−1	0	0

The two dimensionless products are identified to be

$$\prod_1 = \frac{M}{\eta\Omega R^3} \quad \text{and} \quad \prod_2 = C\frac{\rho\,\Omega R^2}{\eta}. \tag{B.69}$$

Equation (B.68) can thus be written in the form

$$\frac{M}{\eta\Omega R^3} = f\left(C\frac{\rho\,\Omega R^2}{\eta}\right). \tag{B.70}$$

The product $\prod_2$ corresponds to the Reynolds number, so that for Re $\to 0$ one obtains

$$\frac{M}{\eta\Omega R^3} = \text{const.} \tag{B.71}$$

The same result is found, when not accounting for the constant C, implying that Newton's law is disregarded and therefore momentum is neglected. For finite, non-zero Reynolds number, it is helpful to choose the products such that the viscosity occurs only in one product, hence

$$\prod_1^* = \prod_1 \prod_2^{-1} = \frac{1}{C}\frac{M}{\rho\,(\Omega R)^2 R^3} \tag{B.72}$$

or in a form equivalent to Eq. (B.70)

$$\frac{M}{\rho(\Omega R)^2 R^3} = f\left(\frac{\rho\,\Omega R^2}{\eta}\right). \tag{B.73}$$

Problem A4.1 It is sought for the resistance W of a sphere in the form

$$W = f(g, v, \sigma, h, d, U, \rho). \tag{B.74}$$

In the $[LMT]$-system the dimension matrix results in

	W	g	v	σ	h	d	U	ρ
	k_1	k_2	k_3	k_4	k_5	k_6	k_7	k_8
L	1	1	2	0	1	1	1	−3
M	1	0	0	1	0	0	0	1
T	−2	−2	−1	−2	0	0	−1	0

The quantities that are to occur linearly in the dimensionless products are put to the front columns. The rank r of the dimension matrix is $r = 3$ so that $d = n - r = 5$ dimensionless products have to be determined. For the exponent of the basic variables to vanish the relation

$$\sum_{j=1}^{n} a_{ij} k_j = 0 \tag{B.75}$$

must hold. In the present case this yields the system of equations

$$\begin{bmatrix} 1 & 1 & 2 & 0 & 1 & 1 & 1 & -3 \\ 1 & 0 & 0 & 1 & 0 & 0 & 0 & 1 \\ -2 & -2 & -1 & -2 & 0 & 0 & -1 & 0 \end{bmatrix} \begin{bmatrix} k_1 \\ k_2 \\ k_3 \\ k_4 \\ k_5 \\ k_6 \\ k_7 \\ k_8 \end{bmatrix} = 0 \tag{B.76}$$

and accordingly

$$\begin{bmatrix} 1 & 1 & 2 & 0 & 1 \\ 1 & 0 & 0 & 1 & 0 \\ -2 & -2 & -1 & -2 & 0 \end{bmatrix} \begin{bmatrix} k_1 \\ k_2 \\ k_3 \\ k_4 \\ k_5 \end{bmatrix} + \begin{bmatrix} 1 & 1 & -3 \\ 0 & 0 & 1 \\ 0 & -1 & 0 \end{bmatrix} \begin{bmatrix} k_6 \\ k_7 \\ k_8 \end{bmatrix} = 0 \tag{B.77}$$

respectively after transformation

$$\begin{bmatrix} 1 & 1 & -3 \\ 0 & 0 & 1 \\ 0 & -1 & 0 \end{bmatrix} \begin{bmatrix} k_6 \\ k_7 \\ k_8 \end{bmatrix} = - \begin{bmatrix} 1 & 1 & 2 & 0 & 1 \\ 1 & 0 & 0 & 1 & 0 \\ -2 & -2 & -1 & -2 & 0 \end{bmatrix} \begin{bmatrix} k_1 \\ k_2 \\ k_3 \\ k_4 \\ k_5 \end{bmatrix} \tag{B.78}$$

and

$$\begin{bmatrix} k_6 \\ k_7 \\ k_8 \end{bmatrix} = - \begin{bmatrix} 1 & 3 & 1 \\ 0 & 0 & -1 \\ 0 & 1 & 0 \end{bmatrix} \begin{bmatrix} 1 & 1 & 2 & 0 & 1 \\ 1 & 0 & 0 & 1 & 0 \\ -2 & -2 & -1 & -2 & 0 \end{bmatrix} \begin{bmatrix} k_1 \\ k_2 \\ k_3 \\ k_4 \\ k_5 \end{bmatrix}. \tag{B.79}$$

Thus, we obtain

$$\begin{bmatrix} k_6 \\ k_7 \\ k_8 \end{bmatrix} = - \begin{bmatrix} 2 & -1 & 1 & 1 & 1 \\ 2 & 2 & 1 & 2 & 0 \\ 1 & 0 & 0 & 1 & 0 \end{bmatrix} \begin{bmatrix} k_1 \\ k_2 \\ k_3 \\ k_4 \\ k_5 \end{bmatrix}. \tag{B.80}$$

This results in a set of $d = n - r = 5$ linearly independent solution vectors $k_{(i),j}, (i = 1\ldots d, j = 1\ldots n)$ by assigning the $k_{(i),j}, (i,j = 1\ldots d)$ on the right side of Eq. (B.80) to unit vectors according to

$$\begin{bmatrix} k_{(1),1} & k_{(1),2} & k_{(1),3} & k_{(1),4} & k_{(1),5} \\ k_{(2),1} & k_{(2),2} & k_{(2),3} & k_{(2),4} & k_{(2),5} \\ k_{(3),1} & k_{(3),2} & k_{(3),3} & k_{(3),4} & k_{(3),5} \\ k_{(4),1} & k_{(4),2} & k_{(4),3} & k_{(4),4} & k_{(4),5} \\ k_{(5),1} & k_{(5),2} & k_{(5),3} & k_{(5),4} & k_{(5),5} \end{bmatrix} = \begin{bmatrix} 1 & 0 & 0 & 0 & 0 \\ 0 & 1 & 0 & 0 & 0 \\ 0 & 0 & 1 & 0 & 0 \\ 0 & 0 & 0 & 1 & 0 \\ 0 & 0 & 0 & 0 & 1 \end{bmatrix}. \tag{B.81}$$

Substituting Eq. (B.81) into Eq. (B.80), allows to determine the $k_{(i),1}, k_{(i),2}, k_{(i),3}, (i = 1\ldots d)$ according to the following table.

	k_1	k_2	k_3	k_4	k_5	k_6	k_7	k_8
Π_1	1	0	0	0	0	−2	−2	−1
Π_2	0	1	0	0	0	1	−2	0
Π_3	0	0	1	0	0	−1	−1	0
Π_4	0	0	0	1	0	−1	−2	−1
Π_5	0	0	0	0	1	−1	0	0

The dimensionless products then follow from

$$\prod_i = W^{k_{(i),1}}\, g^{k_{(i),2}}\, \nu^{k_{(i),3}}\, \sigma^{k_{(i),4}}\, h^{k_{(i),5}}\, d^{k_{(i),6}}\, U^{k_{(i),7}}\, \rho^{k_{(i),8}}, \quad (i = 1\ldots d) \tag{B.82}$$

to

$$\prod_1 = \frac{W}{d^2\, U^2\, \rho} = c_w, \tag{B.83}$$

$$\prod_2 = \frac{g\, d}{U^2} = \frac{1}{Fr}, \tag{B.84}$$

$$\prod_3 = \frac{\nu}{d\, U} = \frac{1}{\mathrm{Re}}, \tag{B.85}$$

$$\prod_4 = \frac{\sigma}{d\, U^2 \rho} = We, \tag{B.86}$$

and

$$\prod_5 = \frac{h}{d}. \tag{B.87}$$

Thus, Eq. (B.74) can be written in the equivalent form

$$c_w = f(\mathrm{Re}, \mathrm{Fr}, \mathrm{We}, h/d). \tag{B.88}$$

In most practical applications, the Laplace length $a \equiv \sqrt{\sigma/\rho\, g}$ is much smaller than the sphere diameter, thus

$$\frac{a}{d} \ll 1. \tag{B.89}$$

Then, the surface tension becomes insignificant for the problem and the Weber number may be neglected. Formally, the products are combined to

$$\prod_4^* = \sqrt{\mathrm{Fr}}\sqrt{\mathrm{We}} = \frac{U}{\sqrt{g\, d}}\sqrt{\frac{\sigma}{\rho\, d}}\frac{1}{U} = \frac{a}{d} \tag{B.90}$$

with $\prod_4^* \ll 1$ resulting in

$$c_w = f(\mathrm{Re}, \mathrm{Fr}, h/d). \tag{B.91}$$

Problem A5.1 In consideration of geometric similarity the buckling load is only a function of the elasticity modulus E and a characteristic length L, thus

$$P = f(E, L). \tag{B.92}$$

The corresponding dimension matrix reads

	P	E	L
L	1	−1	1
M	1	1	0
T	−2	−2	0

and the only dimensionless product is identified to be

$$\frac{P}{E\,L^2} = \text{const.} \tag{B.93}$$

To achieve complete similarity, all dimensionless products for model and full-scale must be equal. The ratios of the physical quantities in the model p'_j to those of the full-scale p_j determine the scale factors M_j, i.e.

$$p'_j = M_j p_j. \tag{B.94}$$

In the present case, the dimensionless quantity given by Eq. (B.93) for model and full-scale must be equal, i.e.

$$\frac{P}{E\,L^2} = \frac{P'}{E'\,L'^2} \tag{B.95}$$

or

$$\frac{P}{P'} = \frac{E}{E'}\left(\frac{L}{L'}\right)^2 \tag{B.96}$$

resulting in the solution

$$P = 7.5 \times 10^6\ \text{N}. \tag{B.97}$$

Problem A5.2 A cantilever that is anchored at one end and carries a load P at the other end is considered and it is sought for the deflection of the beam. The deflection of the beam w depends on the force P, the beam length L, the density ρ and the gravitational acceleration g so that the relation is of the form

$$w = f(L, P, E, \rho, g). \tag{B.98}$$

The corresponding dimension matrix reads

	w	L	P	E	ρ	g
L	1	1	1	−1	−3	1
M	0	0	1	1	1	0
T	0	0	−2	−2	0	−2

and after transformation

	w/L	L	P/E	E	ρ/E	g
L	0	1	2	−1	−2	1
M	0	0	0	1	0	0
T	0	0	0	−2	2	−2

and

	w/L	L	$P/(E\,L^2)$	E	$\rho\,g\,L/E$	g
L	0	1	0	−1	0	1
M	0	0	0	1	0	0
T	0	0	0	−2	0	−2

three dimensionless product are obtained and Eq. (B.98) is rewritten to

$$\frac{w}{L} = f\left(\frac{P}{E\,L^2}, \frac{\rho\,g\,L}{E}\right). \tag{B.99}$$

In order for the model and full-scale are loaded in a physically similar way, the dimensionless products for both versions must be equal, i.e.

$$\frac{P}{E\,L^2} = \frac{P'}{E'\,L'^2} \quad \text{and} \quad \frac{\rho\,g\,L}{E} = \frac{\rho'\,g\,L'}{E'}, \tag{B.100}$$

or

$$\frac{P'}{P} = \frac{E'}{E}\left(\frac{L'}{L}\right)^2 \quad \text{and} \quad \frac{\rho'}{\rho} = \frac{E'}{E}\frac{L}{L'}. \tag{B.101}$$

Accordingly the scale factors are related by

$$M_P = M_E M_L^2 \quad \text{and} \quad M_\rho = \frac{M_E}{M_L} \tag{B.102}$$

or

$$M_P = M_E M_L^2 = M_\rho M_L^3. \tag{B.103}$$

From Eq. (B.102) it follows that for the same material ($M_E = M_\rho = 1$) model and full-scale must be the same, if the deflection as a result of its own weight is considered. If the deformation caused by its own weight can be neglected over the single force deformation, the force in the model is smaller than that for the full scale by a factor of M_L^2.

To determine the stresses σ, a relation of the form

$$\sigma = f(L, P, E, \rho, g), \tag{B.104}$$

is examined, corresponding to the non-dimensional formulation

$$\frac{\sigma L^2}{P} = f\left(\frac{P}{E L^2}, \frac{\rho g L}{E}\right). \tag{B.105}$$

The demand for the equality of the dimensionless products for model and full scale yields the relations for the scale factors to

$$M_\sigma = \frac{M_P}{M_L^2}, \quad M_P = M_E M_L^2 \quad \text{and} \quad M_\rho = \frac{M_E}{M_L}, \tag{B.106}$$

respectively

$$M_\sigma = M_E \quad \text{and} \quad M_P = M_E M_L^2 = M_\rho M_L^3. \tag{B.107}$$

It is again found that when using the same material, model and full-scale must be the same if stresses caused by the own weight cannot be neglected. In case that the own weight may be neglected over the added loading it is found that the stresses in the model and in the full scale are equal.

Problem A5.3 The temperature of the roast beef piece ϑ depends on the initial temperature ϑ_0 at which it has been pushed into the oven, the oven temperature ϑ_∞, the time t, the thermal conductivity k, the heat capacity c, the density ρ, and a characteristic length L. We are therefore looking for a functional relation of the form

$$\vartheta = f(\vartheta_0, \vartheta_\infty, t, k, \rho, c, L). \tag{B.108}$$

Since heat conduction problems are linear with regard to the temperature and therefore only temperature differences are significant, we write Eq. (B.108) in the form

$$\vartheta - \vartheta_0 = (\vartheta - \vartheta_\infty) \quad f(t, k, \rho, c, L). \tag{B.109}$$

So that the dimension matrix yields

	t	k	ρ	c	L
L	0	1	−3	2	1
M	0	1	1	0	0
T	1	−3	0	−2	0
θ	0	−1	0	−1	0

and after transformation

	t	$k/(\rho c)$	ρ	c	L
L	0	2	−3	2	1
M	0	0	1	0	0
T	1	−1	0	−2	0
θ	0	0	0	−1	0

The only dimensionless product then becomes

$$\frac{t}{L^2} \frac{k}{\rho c} = \text{const.} \tag{B.110}$$

or

$$t = \text{const.} L^2 \frac{\rho c}{k}. \tag{B.111}$$

The mass of the roast beef piece is

$$m \propto \rho L^3 \quad \text{thus} \quad L \propto \left(\frac{m}{\rho}\right)^{1/3}, \tag{B.112}$$

so that accordingly Eq. (B.111) becomes

$$t = \text{const.} \frac{\rho c}{k} \left(\frac{m}{\rho}\right)^{2/3} = \text{const.} \frac{c}{k} \rho^{1/3} m^{2/3}. \tag{B.113}$$

The time t' that the roast beef piece with $m' = 2\,kg$ must fry, can then be determined from the roasting time t of the lighter piece with $m = 1\ kg$

$$\frac{t'}{t} = \left(\frac{m'}{m}\right)^{2/3} \quad \text{or} \quad t' = t\left(\frac{m'}{m}\right)^{2/3} = 95\ \text{min}. \tag{B.114}$$

Problem A6.1 The sought circumferential speed is a function of the form

$$v = v(r, t, \Gamma, \eta, \rho). \tag{B.115}$$

(a) The dimension matrix reads

	v	r	t	Γ	η	ρ
L	1	1	0	2	−1	−3
M	0	0	0	0	1	1
T	−1	0	1	−1	−1	0

and may be transformed to

	$v\,r$	r	t	Γ	$\eta/\rho = \nu$	ρ
L	2	1	0	2	2	−3
M	0	0	0	0	0	1
T	−1	0	1	−1	−1	0

and further to

	$v\,r/\Gamma$	$r/\sqrt{\nu\,t}$	t	Γ/ν	ν
L	0	0	0	0	2
M	0	0	0	0	0
T	0	0	1	0	−1

So that the nondimensional products result to

$$\Pi_1 = \frac{v\,r}{\Gamma}\ , \quad \Pi_2 = \frac{r}{\sqrt{\nu\,t}} \quad \text{and} \quad \Pi_3 = \frac{\Gamma}{\nu}. \tag{B.116}$$

With Eq. (B.116) the relation

$$v = f(r, t, \Gamma, \eta, \rho) \tag{B.117}$$

takes the form

$$\frac{v\,r}{\Gamma} = f\left(\frac{r}{\sqrt{\nu\,t}}, \frac{\Gamma}{\nu}\right). \tag{B.118}$$

Accordingly, a similarity variable is introduced by

$$\eta = \frac{r}{\sqrt{\nu\,t}}. \tag{B.119}$$

and we set

$$\frac{v\,r}{\Gamma} = F(\eta). \tag{B.120}$$

The initial and boundary conditions Eqs. (A.1), (A.2) and (A.3) then result in

$$F(0) = 0 \quad \text{and} \quad F(\infty) = \frac{1}{2\,\pi}. \tag{B.121}$$

Introducing the similarity variable

$$\frac{\partial \eta}{\partial t} = -\frac{\eta}{2\,t} \quad \text{and} \quad \frac{\partial \eta}{\partial r} = \frac{1}{\sqrt{\nu\,t}} \tag{B.122}$$

into the differential equation

$$\frac{\partial v}{\partial t} = \nu\left(\frac{\partial^2 v}{\partial r^2} + \frac{1}{r}\frac{\partial v}{\partial r} - \frac{v}{r^2}\right) \tag{B.123}$$

results in the following ordinary differential equation

$$-\frac{\eta^3}{2} F' = -\eta\, F' + \eta^2\, F'' \tag{B.124}$$

that may be simplified to

$$\frac{F''}{F'} = -\frac{\eta}{2} + \frac{1}{\eta}. \tag{B.125}$$

This equation can be solved by using the method of separation of variables. Integration yields

$$\ln(F') = -\frac{\eta^2}{4} + \ln(\eta) + \ln(c_1) \tag{B.126}$$

with the integration constant c_1. Equation (B.126) may be simplified to

$$\ln\left(\frac{F'}{c_1\eta}\right) = -\frac{\eta^2}{4} \tag{B.127}$$

or

$$F' = c_1\eta \exp\left(-\frac{\eta^2}{4}\right). \tag{B.128}$$

Integration of Eq. (B.128) results to

$$F = 2\,c_1 \exp\left(-\frac{\eta^2}{4}\right) + c_2. \tag{B.129}$$

Using the initial and boundary conditions of Eq. (B.121) the constants are determined to be

$$c_2 = \frac{1}{2\,\pi} \quad \text{and} \quad 2\,c_1 = -c_2 \tag{B.130}$$

so that the solution becomes

$$F = \frac{1}{2\,\pi}\left(1 - \exp\left(-\frac{\eta^2}{4}\right)\right) \tag{B.131}$$

respectively

$$v = \frac{\Gamma}{2\,\pi\,r}\left(1 - \exp\left(-\frac{\eta^2}{4}\right)\right). \tag{B.132}$$

References

Barenblatt GI (1987) Dimensional analysis. Gordon & Breach, New York

Bridgman P (1922) Dimensional analysis. Yale University Press, New Haven

Bronstein IN, Semendjajew KA (1981) Taschenbuch der Mathematik. Verlag Harri Deutsch, Frankfurt

Buckingham E (1914) On physically similar systems. Illustrations of the use of dimensional equations. Phys Rev 4:345–376

Cordier O (1953) Ähnlichkeitsbedingungen für Strömungsmaschinen. BWK 5:337–340

Cordier O (1955) Probleme der Strömungstechnik im Maschinenbau: Ähnlichkeitsbetrachtungen für Strömungsmaschinen. VDI-Berichte 3:85–88

Cowling TG (1957) Magnetohydrodynamics. Interscience, New York

Crane LJ (1970) Flow past a stretching plane. Zeitschrift für angewandte Mathematik und Physik 21:645–647

DIN 1305 (1974) Normen für Größen und Einheiten in Naturwissenschaft und Technik. Beuth, Berlin

Dixon SL (1978) Fluid mechanics, thermodynamics of turbomachinery. Pergamon Press, Oxford

Eckert EGR, Drake RM (1972) Analysis of heat and mass transfer. McGraw Hill, New York

Eirich-Pelerie R (1950) Dimensional analysis. Lausanne, Switzerland

Epshteyn LA (1977) Methods of the dimensional analysis and similarity theory in problems of ship hydromechanics. Department of the Naval Intelligence Support Center, Washington D.C.

Galileo G (1953) Dialogues concerning two new sciences. 1638, cited in Focken CM, Dimensional methods and their applications. Arnold, London

Görtler H (1975) Dimensionsanalyse. Springer, Berlin

Greenhill AG (1881) Determination of the greatest height consistent with stability that a vertical pole or mast can be made, and of the greatest height to which a tree of given proportion can grow. Proc Camb Phil Soc 4:65–73

Hansen AG (1964) Similarity analyses of boundary value problems in Engineering. Prentice Hall, Upper Saddle River

Horlock JH (1958) Axial flow compressors—fluid mechanics and thermodynamics. Butterworths, London

Horlock JH (1973) Axial flow turbines—fluid mechanics and thermodynamics. Krieger Publishing Company, Malabar

Huntley HE (1953) Dimensional analysis. McDonald, London

Kays WM, Crawford ME, Weigand B (2004) Convective heat and mass transfer. Mc. Graw-Hill Inc., New York

Landau LD, Lifshitz EM (1987) Fluid mechanics. Pergamon Press, Oxford

Langhaar HL (1951) Dimensional analysis and theory of models. Wiley, New York

Liebe R (2004) Applications of the finite vortex model. In: Proceedings of the W.I.T. 2nd international conference on "design & nature", Rhodes, pp 371–380

V. Simon et al., *Dimensional Analysis for Engineers*,
Mathematical Engineering, DOI 10.1007/978-3-319-52028-5

Liebe W, Liebe R (2002) A general finite vortex model to describe unsteady aerodynamics in nature. In: Proceedings of the W.I.T. 1st international conference on "design & nature", Udine, pp 337–346

Lighthill J (1975) Mathematical biofluid dynamics. Society of Industrial and Applied Mathematics (SIAM), Philadelphia

Maxwell JC (1871) On the mathematical classification of physical quantities. Proc London Math Soc 3:224

McMahon TA (1971) Rowing: a similarity analysis. Science 173:349–351

McMahon TA (1975) The mechanical design of trees. Sci Am 233:92–102

McMahon TA, Bonner JT (1985) Form und Leben: Konstruktionen vom Reißbrett der Natur. Spektrum-der-Wissenschaft-Verlagsgesellschaft, Heidelberg

Panton RL (1984) Incompressibile flow. Wiley, New York

Pedley TJ (1974) Scale effects in animal locomotion. Academic Press, New York

Rayleigh Lord JWS (1915) The principles of similitude. Nature 95:66–68

Schlichting H (1979) Boundary layer theory. McGraw Hill, New York

Sedov LI (1959) Similarity and dimensional methods in mechanics. Academic Press, New York

Spurk JH (1992) Dimensionsanalyse in der Strömungslehre. Springer, Berlin

Spurk JH, Aksel N (2010) Strömungslehre, 8th edn. Springer, Berlin

Simon V, Jenkins JT (1994) On the vertical structure of dilute planetary rings. Icarus 110:109–116

Stahl WR (1962) Similarity and dimensional methods in biology. Science 137:205–212

Szabo I (1975) Einführung in die Technische Mechanik. Springer, Berlin

Taylor GI (1950) The formation of a blast wave by a very intense explosion. Proc. Royal Soc. A 201:159–186

Taylor ES (1974) Dimensional analysis for engineers. Clarendon Press, Oxford

Tennekes H, Lumley JL (1972) A first course in turbulence. The MIT Press, Cambridge

Thompson DW (1917) On growth and form. University Press, Cambridge

Vincenti WG, Kruger CH (1986) Introduction to physical gas dynamics. Krieger Publishing Company, Malabar, Florida

Wang CY (1984) The three-dimensional flow due to a stretching flat surface. Phys Fluids 27:1915–1917

Wang CY (1991) Exact solutions of the steady-state Navier-Stokes equations. Ann Rev Fluid Mech 23:159–177

Weigand B (2015) Analytical methods for heat transfer and fluid flow problems, 2nd edn. Springer, Berlin

Weigand B, Simon V (2007) Laws of similarity in fluid mechanics. In: Liebe R (ed) Flow phenomena in nature, vol 1. WIT Press, Ashurt Lodge, pp 20–35

Weigand B, Köhler J, von Wolfersdorf J (2016) Thermodynamik kompakt, 4th edn. Springer, Berlin

Wieghardt K (1952) Über einige Versuche an Strömungen in Sand. Ing.-Archiv **20**:109–115

Zierep J (1972) Ähnlichkeitsgesetze und Modellregeln der Strömungslehre. G. Braun, Karlsruhe

Zlokarnik M (1998) Dimensional analysis: scale-up. In: Proceedings of international workshop on similarity methods, Stuttgart, pp 1–56

Index

V. Simon et al., *Dimensional Analysis for Engineers*,
Mathematical Engineering, DOI 10.1007/978-3-319-52028-5

Lightning Source UK Ltd.
Milton Keynes UK
UKHW02n1138221018
330975UK00004B/83/P